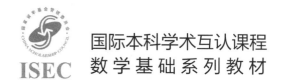

国际本科学术互认课程
数学基础系列教材

微积分 I（双语版）

程晓亮　王洋　华志强　盛婷婷　侯颖 ◎编著

图书在版编目(CIP)数据

微积分：双语版. Ⅰ / 程晓亮等编著. —北京：北京大学出版社，2017.9
（国际本科学术互认课程·数学基础系列教材）
ISBN 978-7-301-28630-2

Ⅰ. ①微…　Ⅱ. ①程…　Ⅲ. ①微积分—高等学校—教材　Ⅳ. ①O172

中国版本图书馆 CIP 数据核字（2017）第 199356 号

书　　　名	微积分Ⅰ（双语版）
	WEIJIFEN Ⅰ
著作责任者	程晓亮　王　洋　华志强　盛婷婷　侯　颖　编著
责 任 编 辑	曾婉婷
标 准 书 号	ISBN 978-7-301-28630-2
出 版 发 行	北京大学出版社
地　　　址	北京市海淀区成府路 205 号　100871
网　　　址	http://www.pup.cn　新浪微博：@北京大学出版社
电 子 信 箱	zpup@pup.cn
电　　　话	邮购部 62752015　发行部 62750672　编辑部 62767347
印 刷 者	北京宏伟双华印刷有限公司
经 销 者	新华书店
	889 毫米×1194 毫米　16 开本　11.5 印张　338 千字
	2017 年 9 月第 1 版　2024 年 5 月第 7 次印刷
定　　　价	56.00 元

未经许可，不得以任何方式复制或抄袭本书之部分或全部内容。
版权所有，侵权必究
举报电话：010-62752024　电子信箱：fd@pup.pku.edu.cn
图书如有印装质量问题，请与出版部联系，电话：010-62756370

内容简介

本书是根据"国际本科学术互认课程"(ISEC)项目对高等数学系列课程的要求,同时结合 ISEC 项目培养模式进行编写的"微积分"双语教材.全书共分 6 章,内容包括:函数、极限、导数和微分、导数的应用、不定积分、定积分等.在内容选择上,既考虑到 ISEC 学生未来学习和发展的需要,又兼顾学生数学学习的实际情况,以适用、够用为原则,切合学生实际,在体系完整的基础上对通常的"微积分"课程内容进行适当的调整,注重明晰数学思想与方法,强调数学知识的应用;在内容阐述上,尽量以案例模式引入,由浅入深,由易到难,循序渐进地加以展开,并且尽量使重点突出,难点分散,便于学生对知识的理解和掌握;在内容呈现上,以英文和中文两种文字进行编写,分左、右栏对应呈现,方便学生学习与理解.

本书既可作为 ISEC 项目培养模式下"微积分"课程的教材,也可作为普通高等院校"微积分"课程的教学参考书,特别是以英文和中文两种语言学习和理解"微积分"的参考资料.

为方便教学,作者为任课教师提供相关的电子资源,具体事宜可通过电子邮件与作者联系,邮箱地址:chengxiaoliang92@163.com.

序　　言

时值"国际本科学术互认课程·数学基础系列教材"之一《微积分Ⅰ(双语版)》面世之际,本人在此向程晓亮老师和参加这套教材编写的各位 ISEC 教师表示热烈祝贺.

"国际本科学术互认课程"(International Scholarly Exchange Curriculum,简称 ISEC)项目,是国家留学基金管理委员会主持的、面向国内地方本科院校的教学改革项目.该项目致力于建设集国际化课程、国际化师资、国际教育资源于一体的国际化教育教学工作平台,并依托该平台,将具有国际先进水平的教学理念、教学思想和教学方法融入教师的教学实践,推动地方高校的教学改革.

本人深入参与了 ISEC 项目的两个基本环节:一是教师的课堂教学设计;二是教师和学生的明辨性思维训练.近距离观察了 ISEC 课程的教师和学生,有如下基本印象:教师培训成效显著.ISEC 教师对现代教育教学的思想和方法有相当程度的理解,对教师培训的内容有相当好的回应和反馈,对明辨性思维有相当程度的认知;ISEC 学生显示出不错的灵气和悟性.

ISEC 项目已经有坚实的基础,还有很大的发展空间,会有光明的发展前景.

EMI(English Media Instruction)是 ISEC 课程教学的特点之一,目前适合 EMI 教学的高等数学系列课程教材处于空白.为了适应 ISEC 课程教学的需要,解决 ISEC 学生高等数学的学习困难,在国家留学基金管理委员会 ISEC 项目办公室的关心和支持下,由程晓亮老师牵头,ISEC 项目院校吉林师范大学、内蒙古民族大学、贵州财经大学、贵阳学院、包头师范学院和赤峰学院等院校的ISEC 教师参与,组织编写了一套英中对照教材——"国际本科学术互认课程·数学基础系列教材",包括《微积分Ⅰ(双语版)》《微积分Ⅰ习题解析(双语版)》《微积分Ⅱ(双语版)》《微积分Ⅱ习题解析(双语版)》《线性代数(双语版)》《线性代数习题解析(双语版)》《概率论与数理统计(双语版)》和《概率论与数理统计习题解析(双语版)》.

这些 ISEC 教师因参与 ISEC 项目,了解了现代教育教学的思想和理念,掌握了课堂教学设计的方法和技巧,养成了明辨性思维的意识和习惯,并且将在 ISEC 项目中学习到的理念、思想和方法与自己的教学实践相结合,编写了这套双语教材,为推进 EMI 教学提供了有力的支撑.

已编写完成的《微积分Ⅰ(双语版)》,其语言显示出如下特点:教材中的英语不仅语言流畅,用词准确,而且充分兼顾了西方读者的思维习惯;教材中的中文与英文,不仅在空间形式上而且在内涵上形成了准确对应.更令人印象深刻的是,中文表述完全符合中国读者的思维习惯.可以说,这部双语教材有效地平衡了东、西方读者在思维习惯上的差异.

ISEC 项目高度重视教师和学生的明辨性思维素质的养成.这一意图在《微积分Ⅰ(双语版)》中得到不折不扣的贯彻:从基本概念的抽象到基本定理的证明,从基本思想的发展到理论体系的构建,无不体现出明辨性思维的特质.可以说,作者将明辨性思维有效地融入了教材的每一个章节甚至字里行间.因此,该教材是一部优秀的双语教材,也是养成学生明辨性思维素质的好教材.

另外,《微积分Ⅰ(双语版)》充分顾及学生认知发展的基本规律.德国伟大的数学家希尔伯特时常告诫自己的学生"从鲜活的案例开始",肯定了"鲜活的案例"在抽象数学概念、生成数学思想方面的重大作用.这部教材自觉遵循了希尔伯特的忠告,其案例的选择、内容的展开、理论的陈述,都遵循了由浅入深、由简单到复杂、由具体到抽象、由特殊到一般的基本原则.

随着ISEC项目的推进,我们期待能有更多这样的好教材面世.

殷雅俊
ISEC项目专家
清华大学教授
2017年6月于北京

前　　言

　　党的二十大报告对实施科教兴国战略、强化现代化建设人才支撑作出重大部署,明确指出:"教育、科技、人才是全面建设社会主义现代化国家的基础性、战略性支撑". 青年强,则国家强. 广大教师深受鼓舞,更要勇担"为党育人,为国育才,全面提高人才自主培养质量"的重任,迎来一个大有可为的新时代. 在多种形式的人才培养途径中,要提升学生的国际视野,同时坚守中华文化立场,深化文明交流互鉴. 这也正是编写出版"国际本科学术互认课程·数学基础系列教材"这套中英双语对照教材的初衷.

　　本书是"国际本科学术互认课程·数学基础系列教材"之一,它紧密结合国际本科学术互认课程(International Scholarly Exchange Curriculum,简称 ISEC)对教学的要求,强调学习知识与训练思维的统一,强调教学理念与方法的统一,强调学习过程与学生能力提升的统一. 在内容选择上,既考虑到 ISEC 学生未来学习和发展的需要,又兼顾学生数学学习的实际情况,以适用、够用为原则,切合学生实际,在体系完整的基础上对通常的"微积分"课程内容进行适当的调整,注重明晰数学思想与方法,强调数学知识的应用;在内容阐述上,尽量以案例模式引入,由浅入深,由易到难,循序渐进地加以展开,并且尽量使重点突出,难点分散,便于学生对知识的理解和掌握;在内容呈现上,以英文和中文两种文字进行编写,分左、右栏对应呈现,方便学生学习与理解.

　　本书作者都是经过多次 ISEC 教师岗前培训和专题培训的教师,并多次承担 ISEC 课程"微积分"的教学工作. 正是在教学过程中,我们发现国内适应国际化教育教学需要的"微积分"双语教材匮乏. 我们曾试图直接采用英文原版的"微积分"教材. 但是,由于学生英文水平的限制以及以前没有双语学习的基础,特别是对"微积分"中涉及思想和方法的内容,学生把握起来比较困难. 所以说,直接采用英文原版教材在某种程度上是不合适的. 就"微积分"而言,国内有很多优秀的教材. 然而,根据 ISEC 课程对学生发展的目标要求,是不宜采用中文教材的. 正是在这样的背景下,我们结合多轮"微积分"教学经验,精心选材与设计,撰写了这部"微积分"双语教材.

　　全书由程晓亮、王洋撰写,参与编写、审阅、修改工作的还有华志强、盛婷婷、侯颖.

　　在本书的编写过程中,我们得到了国家留学基金管理委员会 ISEC 项目办公室的大力支持. 可以说,没有 ISEC 项目办公室的鼓励与支持,就没有这套教材的孕育,更谈不上这套教材的面世. 吉林师范大学教务处各位领导十分关心这套教材的撰写,尤其是李雪飞教授给予了无微不至的关心与大力的支持. 在此,我们表示衷心的感谢.

　　ISEC 项目专家、清华大学教授殷雅俊在百忙之中为这套教材作序. 借助 ISEC 教师岗前培训和专题培训,我们多次得到殷雅俊教授的培训与指导,其内涵丰富、思想深邃,使我们受益匪浅. 在此,特别对殷雅俊教授送上崇高的敬意与万分的感激.

由于我们水平有限,书中难免存在这样或者那样的问题,恳请各位同行和读者批评指正.我们期待本书能不断完善,也期待有更优秀的教材面世.让我们共同努力在 ISEC 平台上成长壮大,为适应国际化的教育发展做好充分的准备.

作 者

2024 年 5 月修订

目 录

Chapter 1　Functions
第 1 章　函数 ················· 1

1.1　Functions and Their Graphs
1.1　函数及其图像 ················· 1

 1. The Domain and the Range of a Function
 1. 函数的定义域和值域 ············ 1

 2. The Graph of a Function
 2. 函数的图像 ··············· 2

 3. The Vertical Line Test for a Function
 3. 函数的垂直线测试 ············ 3

 4. Examples of Functions
 4. 函数的例子 ··············· 4

1.2　The Special Properties of Functions
1.2　函数的特性 ··············· 6

 1. The Boundness of a Function
 1. 函数的有界性 ·············· 6

 2. The Monotonicity of a Function
 2. 函数的单调性 ·············· 7

 3. The Symmetry of a Function
 3. 函数的对称性 ·············· 7

 4. The Periodicity of a Function
 4. 函数的周期性 ·············· 8

1.3　The Operations of Functions
1.3　函数的运算 ··············· 8

 1. The Arithmetic of Functions
 1. 函数的四则运算 ············· 8

 2. The Composition of Functions
 2. 函数的复合 ··············· 9

 3. The Transformations of Functions
 3. 函数的变换 ··············· 10

1.4　Elementary Functions
1.4　初等函数 ················ 11

 1. Basic Elementary Functions
 1. 基本初等函数 ·············· 11

 2. Elementary Functions
 2. 初等函数 ················ 13

Exercises 1
习题 1 ······················ 13

Chapter 2　Limits
第 2 章　极限 ················· 15

2.1　The Limit of a Sequence
2.1　数列的极限 ··············· 15

 1. The Definition of the Convergent Sequence
 1. 收敛数列的定义 ············· 15

 2. The Properties of a Convergent Sequence
 2. 收敛数列的性质 ············· 17

2.2　The Limit of a Function
2.2　函数的极限 ··············· 17

 1. The Limit of a Function as $x \to x_0$
 1. 函数在 $x \to x_0$ 时的极限 ········ 17

 2. One-sided Limits
 2. 单侧极限 ················ 20

 3. The Limit of a Function as $x \to \infty$
 3. 函数在 $x \to \infty$ 时的极限 ········ 22

 4. Infinite Limits
 4. 无穷极限 ················ 23

 5. The Properties of Limits
 5. 极限的性质 ··············· 23

2.3　Limit Laws
2.3　极限运算法则 ·············· 24

2.4　Limit Existence Rules and Two Important Limits
2.4　极限存在准则和两个重要极限 ······ 28

2.5 The Continuity of Functions
2.5 函数的连续性 ………… 32
　1. Continuity at a Point
　1. 在一点处的连续性 ………… 33
　2. Several Common Types of Discontinuities
　2. 间断点的几种常见类型 ………… 34
　3. Continuity on an Interval
　3. 区间上的连续性 ………… 35
　4. The Operations of Continuous Functions
　4. 连续函数的运算 ………… 36
　5. The Properties of Continuous Functions on a Closed Interval
　5. 闭区间上连续函数的性质 ………… 37
2.6 Infinitesimals and Infinitys
2.6 无穷小量和无穷大量 ………… 39
　1. Infinitesimals
　1. 无穷小量 ………… 39
　2. Infinitys
　2. 无穷大量 ………… 40
　3. Compare of Infinitesimals
　3. 无穷小量的比较 ………… 40

Exercises 2
习题 2 ………… 42

Chapter 3　The Derivative and the Differential
第 3 章　导数和微分 ………… 45

3.1 The Concept of the Derivative
3.1 导数的概念 ………… 45
　1. Introducing Examples
　1. 引例 ………… 45
　2. The Derivative of Function at a Point
　2. 函数在一点处的导数 ………… 47
　3. One-sided Derivatives
　3. 单侧导数 ………… 49
　4. The Derivative of a Function
　4. 函数的导数 ………… 50
　5. Relationship Between Differentiability and Continuity
　5. 可导与连续的关系 ………… 52
3.2 The Rules for Finding Derivatives
3.2 求导法则 ………… 53
　1. The Constant Multiple Rule
　1. 常数乘法法则 ………… 53
　2. The Sum Rule
　2. 和法则 ………… 54
　3. The Difference Rule
　3. 差法则 ………… 54
　4. The Product Rule
　4. 乘积法则 ………… 55
　5. The Quotient Rule
　5. 商法则 ………… 55
　6. The Rule for the Derivative of an Inverse Function
　6. 反函数求导法则 ………… 57
　7. The Derivative Formulas of Basic Elementary Functions
　7. 基本初等函数的导数公式 ………… 59
　8. The Chain Rule
　8. 链式法则 ………… 59
3.3 Higher-order Derivatives
3.3 高阶导数 ………… 61
3.4 The Derivatives of Implicit Functions and Functions Determined by Parameter Equations
3.4 隐函数及由参数方程确定的函数的导数 ………… 64
　1. The Derivative of an Implicit Function
　1. 隐函数的导数 ………… 64
　2. The Derivative of a Function Determined by a Parameter Equation
　2. 由参数方程确定的函数的导数 … 67
3.5 The Differential and the Approximation
3.5 微分和近似 ………… 69
　1. The Definition of the Differential
　1. 微分的定义 ………… 70

2. The Rules of the Differential

2. 微分法则 ……………… 71

3. The Differential Formulas of Basic Elementary Functions

3. 基本初等函数的微分公式 ……… 72

4. The Linear Approximation of a Function

4. 函数的线性近似 ……… 72

Exercises 3

习题 3 ………………………… 73

Chapter 4　Applications of the Derivative

第 4 章　导数的应用 ……… 76

4.1　The Mean Value Theorem

4.1　微分中值定理 ……… 76

4.2　The L'Hospital Rule

4.2　洛必达法则 ……… 83

4.3　The Criterion of the Monotonicity of Functions

4.3　函数的单调性判别法 ……… 87

　　1. The First Derivative and Monotonicity

　　1. 函数的一阶导数与单调性 ……… 87

　　2. The Second Derivative and Concavity

　　2. 二阶导数和凹性 ……… 89

4.4　Maxima and Minima

4.4　最大值和最小值 ……… 92

　　1. The Existence Question

　　1. 存在性问题 ……… 92

　　2. Where Do Extreme Values Occur?

　　2. 最值在哪里出现？……… 92

　　3. How to Find Extreme Values?

　　3. 如何求最值？……… 95

4.5　Local Extrema and Local Extrema on Open Intervals

4.5　局部极值与开区间上的局部极值 ……… 97

　　1. Where Do Local Extreme Values Occur?

　　1. 局部极值存在于何处？……… 98

　　2. Extrema on an Open Interval

　　2. 开区间上的最值 ……… 102

4.6　Graphing Functions

4.6　作函数的图像 ……… 103

Exercises 4

习题 4 ………………………… 106

Chapter 5　The Indefinite Integral

第 5 章　不定积分 ……… 108

5.1　The Concept and the Properties of the Indefinite Integral

5.1　不定积分的概念与性质 ……… 108

　　1. The Concepts of the Primitive Function and the Indefinite Integral

　　1. 原函数与不定积分的概念 ……… 108

　　2. Basic Formulas of Integrals

　　2. 基本积分公式 ……… 110

　　3. The Properties of the Indefinite Integral

　　3. 不定积分的性质 ……… 112

5.2　Integration by Substitution

5.2　换元积分法 ……… 114

　　1. The Substitution Rule 1

　　1. 第一换元法 ……… 114

　　2. The Substitution Rule 2

　　2. 第二换元法 ……… 120

5.3　Integration by Parts

5.3　分部积分法 ……… 123

5.4　The Indefinite Integral of the Rational Function

5.4　有理函数的不定积分 ……… 126

　　1. The Indefinite Integral of the Rational Function

　　1. 有理函数的不定积分 ……… 126

　　2. The Indefinite Integral of the Rational Function with Trigonometric Function

　　2. 三角函数有理式的不定积分 ……… 128

　　3. The Indefinite Integral of the Simple Irrational Function

　　3. 简单无理函数的不定积分 ……… 129

Exercises 5

习题 5 ·················· 130

Chapter 6　The Definite Integral
第 6 章　定积分 ·············· 132

6.1　The Concept and the Properties of the Definite Integral

6.1　定积分的概念与性质 ········· 132

　　1. Examples of the Definite Integral

　　1. 定积分问题举例 ········ 132

　　2. The Definition of the Definite Integral

　　2. 定积分的定义 ·········· 133

　　3. The Geometric Significance of the Definite Integral

　　3. 定积分的几何意义 ······ 135

　　4. The Properties of the Definite Integral

　　4. 定积分的性质 ·········· 137

6.2　The Fundamental Formula of Calculus

6.2　微积分基本公式 ·········· 140

　　1. The Function of Integral Upper Limit and Its Derivative

　　1. 积分上限函数及其导数 ······ 140

　　2. The Newton-Leibniz Formula

　　2. 牛顿-莱布尼茨公式 ········ 140

6.3　Definite Integration by Substitution and Parts

6.3　定积分的换元法和分部积分法 ······ 143

　　1. Definite Integration by Substitution

　　1. 定积分的换元法 ··········· 143

　　2. Definite Integration by Parts

　　2. 定积分的分部积分法 ········ 147

6.4　The Improper Integral

6.4　反常积分 ················ 148

　　1. The Improper Integral of Infinite Limit

　　1. 无穷限的反常积分 ········· 148

　　2. The Improper Integral of the Unbounded Function

　　2. 无界函数的反常积分 ········ 151

6.5　Applications of the Definite Integral

6.5　定积分的应用 ·············· 154

　　1. The Infinitesimal Method

　　1. 微元法 ················ 154

　　2. Applications in Geometry

　　2. 在几何中的应用 ·········· 155

　　3. Applications in Economics

　　3. 在经济中的应用 ·········· 166

　　4. Application in Physics

　　4. 在物理中的应用 ·········· 167

Exercises 6

习题 6 ·················· 168

Chapter 1 Functions
第1章 函数

The fundamental objects that we deal with in calculus are functions. In this chapter, we review the basic concepts concerning functions, their graphs, their special properties and the ways of transforming and combining them. We also review the six classes elementary functions and their properties.

在微积分中,我们研究的基本对象是函数.本章我们回顾了函数的相关概念、函数的图形、函数的特性、函数变换和合并的方法.我们也回顾了六大类初等函数以及它们的性质.

1.1 Functions and Their Graphs
1.1 函数及其图像

A function can be represented in different ways: by an equation, in a table, by a graph, or in words.

函数可以用不同的方法表示:公式法,表格法,图像法和描述法.

1. The Domain and The Range of a Function

Definition 1.1 A **function** f from a set D to E is a rule that assigns a unique element $y \in E$ to each element $x \in D$. It is denoted by
$$y = f(x), \quad x \in D.$$

We usually consider functions for which the sets D and E are sets of real numbers. The set D is called the **domain** of the function. The number $f(x)$ is the **value of f at x**. The **range** of f is the set of all possible values of $f(x)$ as x varies throughout the domain D. A symbol x that represents an arbitrary number in the domain of a function f is called an **independent variable**. A symbol y that represents a number in the range of f is called an **dependent variable.**

A function f is like a machine that produces an output value $f(x)$ in its range whenever we feed it an input value x from its domain (Figure 1.1). The function keys on a calculator give an example of a function as a machine.

1. 函数的定义域和值域

定义 1.1 从集合 D 到集合 E 的一个**函数** f,是对 D 中每个元素 $x \in D$ 指定 E 中唯一确定的元素 $y \in E$ 的一种法则,记为
$$y = f(x), \quad x \in D.$$

我们通常研究的函数中,集合 D 和 E 为实数集.称集合 D 为函数 f 的**定义域**,称 $f(x)$ 为函数 f 在**点 x 处的值**.当 x 取遍定义域 D 中所有值时,对应的所有函数值组成的集合称为函数 f 的**值域**.表示函数定义域中任意一个数的变量 x 叫作**自变量**,表示值域中任意一个数的变量 y 称为**因变量**.

函数类似于对每个允许的输入 x 指定一个唯一确定的输出 $f(x)$ 的机器(图1.1).计算器的功能键给出函数作为机器的一个例子.

For instance, the $\sqrt{x}$ key on a calculator gives an output value (the square root) whenever you enter a non-negative number x and press the key $\sqrt{x}$.

例如,当你输入一个非负数 x 并且按下键 $\sqrt{x}$ 时,计算器会给出一个输出值(平方根).

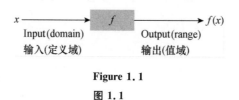

Figure 1.1

图 1.1

2. The Graph of a Function

If f is a function with domain D, its **graph** consists of the points in the Cartesian plane whose coordinates are
$$\{(x,f(x))|x\in D\}.$$

For example, the graph of the function $f(x)=2x-1$ is the set of points with coordinates (x,y) for $y=2x-1$. Its graph is the straight line sketched in Figure 1.2.

2. 函数的图像

假设函数 f 的定义域为 D,它的**图形**由笛卡儿直角坐标平面上所有坐标为
$$\{(x,f(x))|x\in D\}$$
的点集组成.

例如,函数 $f(x)=2x-1$ 的图像指的是坐标平面上所有坐标为 (x,y) ($y=2x-1$) 的点组成的点集. 它的图像是一条直线,如图 1.2 所示.

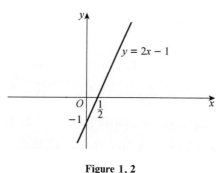

Figure 1.2

图 1.2

The graph of a function f is a useful picture of its behavior. If (x,y) is a point on the graph, then y is the height of the graph above the point x. The height may be positive or negative, depending on the sign of $f(x)$. See Figure 1.3.

函数 f 的图形对于研究函数 f 的性质非常有用. 若 (x,y) 是函数图像上一点,那么 y 就是图像上对应于点 x 的高度,高度可正可负,取决于 $f(x)$ 的符号,见图 1.3.

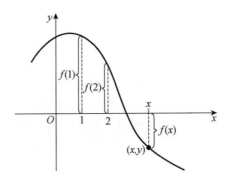

Figure 1.3
图 1.3

3. The Vertical Line Test for a Function

Not every curve in the coordinate plane can be the graph of a function. A function f can have only one value for each x in its domain, so no vertical line can intersect the graph of a function more than once. If a is in the domain of the function f, then the vertical line $x=a$ will intersect the graph of f at the single point $(a, f(a))$.

The Vertical Line Test A curve in the xy-plane is the graph of a function $f(x)$ if and only if no vertical line intersects the curve more than once.

A circle cannot be the graph of a function since some vertical lines intersect the circle twice (Figure 1.4(a)). The circle in Figure 1.4(a), however, does contain the graphs of two functions of x: the upper semicircle defined by the function $f(x)=\sqrt{1-x^2}$ (Figures 1.4(b)), and the lower semicircle defined by the function $f(x)=-\sqrt{1-x^2}$ (Figure 1.4(c)).

3. 函数的垂直线测试

并不是坐标平面上的每一条曲线都可以代表一个函数的图形. 对于定义域中的每一个 x, 函数 f 只能有一个值与其对应, 因此没有一条垂直线可以与一个函数的图像相交多于一次. 如果 a 是函数定义域中的一个点, 那么垂直线 $x=a$ 与函数 f 的图像只能有一个交点 $(a, f(a))$.

垂直线测试 xy 坐标平面上的一条曲线是一个函数 $f(x)$ 的图像, 当且仅当没有任何一条垂直线与该曲线交点多于一个.

一个圆不能成为一个函数的图像, 因为存在垂直线与圆相交两次 (图 1.4(a)). 但是, 图 1.4(a) 中的圆 $x^2+y^2=1$ 包含了两个关于 x 的函数: 一个是上半圆, 函数表达式为 $f(x)=\sqrt{1-x^2}$ (图 1.4(b)); 另一个是下半圆, 函数表达式为 $f(x)=-\sqrt{1-x^2}$ (图 1.4(c)).

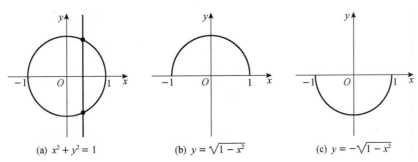

Figure 1.4
图 1.4

4. Examples of Functions

1) The Constant Function

The domain of constant function
$$f(x) = C$$
is $(-\infty, +\infty)$, the range of this function is $\{C\}$, the graph of the constant function is the horizontal line $y=C$ (Figure 1.5).

4. 函数的例子

1) 常函数

常函数
$$f(x) = C$$
的定义域是$(-\infty, +\infty)$，值域是$\{C\}$，它的图像是一条水平直线 $y=C$（图 1.5）.

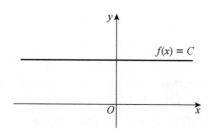

Figure 1.5
图 1.5

2) The Piecewise-Defined Functions

A function described by using different formulas on different parts of its domain is called the **piecewise-defined functions.** One example is the absolute value function
$$|x| = \begin{cases} x, & x \geqslant 0, \\ -x, & x < 0. \end{cases}$$
The right-hand side of the equation means that $f(x)=x$ if $x \geqslant 0$, and $f(x)=-x$ if $x<0$. The graph of this function is given in Figure 1.6.

2) 分段函数

在定义域的不同部分用不同的表达式来定义的函数叫作**分段函数**.

分段函数的一个例子是绝对值函数
$$|x| = \begin{cases} x, & x \geqslant 0, \\ -x, & x < 0. \end{cases}$$
此等式的右端表示，当 $x \geqslant 0$ 时，$f(x)=x$；当 $x<0$ 时，$f(x)=-x$. 该函数的图像如 1.6 所示.

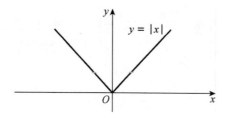

Figure 1.6
图 1.6

Example 1 Suppose that a function is defined by
$$f(x) = \begin{cases} x-1, & x<0, \\ 0, & x=0, \\ x+1, & x>0. \end{cases}$$

例 1 设函数定义如下：
$$f(x) = \begin{cases} x-1, & x<0, \\ 0, & x=0, \\ x+1, & x>0. \end{cases}$$

Evaluate $f(-1), f(0), f(2)$, and sketch the graph.

Solution Since $-1 < 0$, we have
$$f(-1) = -1 - 1 = -2.$$
Since $0 = 0$, we have $f(0) = 0$.
Since $2 > 0$, we have $f(2) = 2 + 1 = 3$.
The graph of this function is given in Figure 1.7.

计算 $f(-1), f(0), f(2)$，并画出图形.

解 由于 $-1 < 0$，我们有
$$f(-1) = -1 - 1 = -2;$$
由于 $0 = 0$，我们有 $f(0) = 0$；
由于 $2 > 0$，我们有 $f(2) = 2 + 1 = 3$.
该函数的图像见图 1.7.

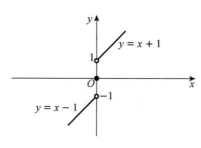

Figure 1.7
图 1.7

3) The Greatest Integer Function

The function whose value at any real number x is the greatest integer less than or equal to x is called the **greatest integer function** or **integer floor function**. It is denoted by $[x]$. Figure 1.8 shows the graph. Observe that
$$[3.6] = 3, \quad [1.9] = 1,$$
$$[0] = 0, \quad [-1.1] = -2.$$

3) 最大取整函数

对于任意实数 x，其值为小于或等于 x 的最大整数的函数称为**最大取整函数**或**整数阶梯函数**，记为 $[x]$. 图 1.8 给出了最大取整函数的图像. 注意到
$$[3.6] = 3, \quad [1.9] = 1,$$
$$[0] = 0, \quad [-1.1] = -2.$$

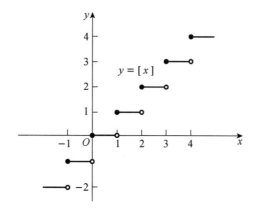

Figure 1.8
图 1.8

4) The Polynomial

A **polynomial** is a function of the form
$$P(x) = a_n x^n + a_{n-1} x^{n-1} + \cdots + a_1 x + a_0,$$

4) 多项式

形如
$$P(x) = a_n x^n + a_{n-1} x^{n-1} + \cdots + a_1 x + a_0,$$

where $a_0, a_1, \cdots, a_n$ are constants. The domain of polynomial is **R**.

5) Ration Function

A **rational function** is a function of the form
$$f(x) = \frac{P(x)}{Q(x)},$$
where $P(x)$ and $Q(x)$ are both polynomials. The domain of $f(x)$ is
$$D = \{x \in \mathbf{R} \mid Q(x) \neq 0\}.$$

的函数叫作**多项式**,其中 $a_0, a_1, \cdots, a_n$ 为常数. 多项式的定义域是 **R**.

5) 有理函数

形如
$$f(x) = \frac{P(x)}{Q(x)}$$
的函数叫作**有理函数**,这里 $P(x)$ 和 $Q(x)$ 都是多项式. 函数 $f(x)$ 的定义域为
$$D = \{x \in \mathbf{R} \mid Q(x) \neq 0\}.$$

1.2 The Special Properties of Functions
1.2 函数的特性

1. The Boundness of a Function

Definition 1.2 A function $f(x)$ defined on some set X is called **bounded** if there exists a real number M such that
$$|f(x)| \leqslant M$$
for all x in X. If M does not exist, $f(x)$ is called **unbounded**.

If $f(x) \leqslant A$ for all x in X, then the function is said to be **bounded above** by X. A is called a **upper bound** of $f(x)$ on X. The graph of the function with upper bound is below the line $y = A$. On the other hand, if $f(x) \geqslant B$ for all x in X, then the function is said to be **bounded below** by X. B is called a **lower bound** of $f(x)$ on X. The graph of the function with lower bound is above the line $y = B$. The graph of the bounded function is between the lines $y = -M$ and $y = M$.

For example, the function $f: \mathbf{R} \to \mathbf{R}$ defined by $f(x) = \sin x$ is bounded, since $|f(x)| \leqslant 1$ for every number $x \in \mathbf{R}$. The function $f(x) = \dfrac{1}{x}$ is unbounded on $(0, 1)$, since there does not exist a constant M, such that $|f(x)| \leqslant M$, but is bounded on $[1, 2]$, because we can take $M = 1$. The function $f(x)$ is bounded on X if and only if it is bounded above and below by X.

1. 函数的有界性

定义 1.2 如果存在实数 M,使得
$$|f(x)| \leqslant M$$
对任意 $x \in X$ 都成立,则称函数 $f(x)$ 在 X 上**有界**. 如果这样的 M 不存在,则称函数 $f(x)$ 在 X 上**无界**.

如果存在实数 A,使得对所有 $x \in X$,有 $f(x) \leqslant A$,则称函数 $f(x)$ 在 X 上**有上界**,而称 A 为函数 $f(x)$ 在 X 上的一个**上界**. 有上界 A 的函数 $y = f(x)$ 的图像在直线 $y = A$ 的下方. 另一方面,如果存在实数 B,使得对所有 $x \in X$,有 $f(x) \geqslant B$,则称函数 $f(x)$ 在 X 上**有下界**,而称 B 为函数 $f(x)$ 在 X 上的一个**下界**. 有下界 B 的函数 $y = f(x)$ 的图像在直线 $y = B$ 的上方. 有界函数的图像在直线 $y = -M$ 和 $y = M$ 之间.

例如,函数 $f(x) = \sin x$ 在 **R** 上是有界的,因为对于任意 $x \in \mathbf{R}$,都有 $|f(x)| \leqslant 1$. 函数 $f(x) = \dfrac{1}{x}$ 在 $(0, 1)$ 上是无界的,因为找不到 M,使得 $|f(x)| \leqslant M$;但函数 $f(x) = \dfrac{1}{x}$ 在 $[1, 2]$ 上是有界的,因为可以取 $M = 1$. 函数 $f(x)$ 在 X 上有界当且仅当它在 X 上既有上界又有下界.

2. The Monotonicity of a Function

Definition 1.3 A function $f(x)$ is called **increasing** on an interval I if
$$f(x_1) < f(x_2)$$
whenever $x_1 < x_2$ in I. It is called the **decreasing** on I if
$$f(x_1) > f(x_2)$$
whenever $x_1 < x_2$ in I.

Increasing functions and decreasing functions are collectively called the **monotonic functions**.

We can see from Figure 1.9 that the function $f(x) = x^2$ is decreasing on the interval $(-\infty, 0]$ and increasing on the interval $[0, +\infty)$.

2. 函数的单调性

定义 1.3 如果对于区间 I 上任意两点 x_1, x_2，当 $x_1 < x_2$ 时，恒有
$$f(x_1) < f(x_2),$$
则称函数 $f(x)$ 在区间 I 上是**单调增加**的；如果对于区间 I 上任意两点 x_1, x_2，当 $x_1 < x_2$ 时，恒有
$$f(x_1) > f(x_2),$$
则称函数 $f(x)$ 在区间 I 上是**单调减少**的.

单调增加和单调减少的函数统称为**单调函数**.

由图 1.9 可知，函数 $f(x) = x^2$ 在区间 $(-\infty, 0]$ 上是单调减少的，在区间 $[0, +\infty)$ 上是单调增加的.

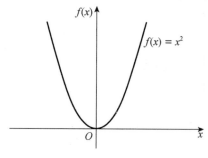

Figure 1.9

图 1.9

3. The Symmetry of a Function

If a function $f(x)$ satisfies
$$f(-x) = f(x)$$
for every number x in its domain D, then $f(x)$ is called an **even function**.

The geometric significance of an even function is that its graph is symmetric with respect to the y-axis (Figure 1.10).

If $f(x)$ satisfies
$$f(-x) = -f(x)$$
for every number x in its domain D, then $f(x)$ is called an **odd function**.

The graph of an odd function is symmetric about the origin (Figure 1.11).

3. 函数的对称性

如果对于函数 $f(x)$ 的定义域 D 中的任意 x，有
$$f(-x) = f(x),$$
则称 $f(x)$ 为**偶函数**.

偶函数的几何意义是，它的图形关于 y 轴对称(图 1.10).

如果对于函数 $f(x)$ 的定义域 D 中的任意 x，有
$$f(-x) = -f(x),$$
则称 $f(x)$ 为**奇函数**.

奇函数的图形关于原点对称(图 1.11).

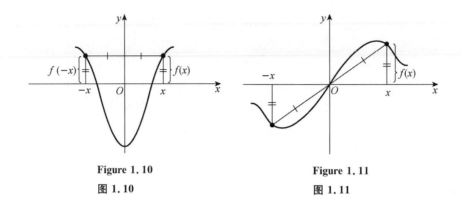

Figure 1.10
图 1.10

Figure 1.11
图 1.11

For example, $y=x^2$, $y=\cos x$ are even functions, $y=x^3$, $y=\sin x$ are odd functions, $y=\sin x+\cos x$ is neither odd or even function.

例如,$y=x^2$,$y=\cos x$ 都是偶函数,$y=x^3$,$y=\sin x$ 都是奇函数,$y=\sin x+\cos x$ 是非奇非偶函数.

4. The Periodicity of a Function

Definition 1.4 A function $f(x)$ is **periodic** if there is a positive number l such that
$$f(x+l)=f(x)$$
for every number x in its domain D. The smallest value of l is the **period** of $f(x)$.

For instance, $\sin x$ and $\cos x$ are common periodic functions, with period 2π.

4. 函数的周期性

定义 1.4 如果存在一个正数 l,使得对于函数 $f(x)$ 定义域 D 中任意 x,都有
$$f(x+l)=f(x),$$
则称 $f(x)$ 为**周期函数**.这样的正数 l 的最小值称为 $f(x)$ 的**周期**.

例如,$\sin x$ 和 $\cos x$ 就是最常见的以 2π 为周期的周期函数.

1.3 The Operations of Functions
1.3 函数的运算

1. The Arithmetic of Functions

Just as two numbers a and b can be added to produce a new number $a+b$, so two functions f and g can be added to produce a new function $f+g$. This is just one of several operations on functions. We can define the following operations of functions: sums, differences, products, and quotients.

The **sum** and **difference functions** are defined by
$$(f+g)(x)=f(x)+g(x),$$

1. 函数的四则运算

正如两个数 a 和 b 可以通过相加得出一个新的数 $a+b$ 一样,两个函数 f,g 也可以通过相加得到一个新函数 $f+g$.这只是函数运算中的一种运算.我们可以定义函数的下列运算:和、差、积和商.

和函数和差函数可以分别定义为
$$(f+g)(x)=f(x)+g(x),$$

$$(f-g)(x) = f(x) - g(x).$$

If the domain of f is A and the domain of g is B, then the domain of $f+g$ is the intersection $A \cap B$ because both f and g have to be defined.

For example, the domain of $f(x) = \sqrt{x+1}$ is $A = [-1, +\infty)$ and the domain of $g(x) = \sqrt{2-x}$ is $B = (-\infty, 2]$, so the domain of $(f+g)(x) = \sqrt{x+1} + \sqrt{2-x}$ is $A \cap B = [-1, 2]$.

Similarly, the **product** and **quotient functions** are defined by
$$(fg)(x) = f(x)g(x),$$
$$\left(\frac{f}{g}\right)(x) = \frac{f(x)}{g(x)}.$$

The domain of fg is $A \cap B$, while the domain of $\frac{f}{g}$ is $\{x \in A \cap B \mid g(x) \neq 0\}$ since we can't divide by 0.

For instance, if $f(x) = \sqrt{2-x}$ and $g(x) = \sqrt{x+1}$, the domain of $\left(\frac{f}{g}\right)(x) = \frac{\sqrt{2-x}}{\sqrt{x+1}}$ is $(-1, 2]$.

2. The Composition of Functions

Composition is another method for combining functions.

Definition 1.5 If f and g are functions, the **composite function** $f \circ g$ is defined by
$$(f \circ g)(x) = f(g(x)).$$

The definition of the composite function implies that $f \circ g$ can be formed when the range of g lies in the domain of f. To find the composite function $(f \circ g)(x)$, first find $g(x)$ and second find $f(g(x))$. To evaluate the composite function $(g \circ f)(x)$, we find $f(x)$ first and then $g(f(x))$. The domain of $g \circ f$ is the set of numbers x in the domain of f such that $f(x)$ lies in the domain of g. The functions $f \circ g$ and $g \circ f$ are usually quite different.

Example 1 If
$$f(x) = \sqrt{x} \quad \text{and} \quad g(x) = 2x^3 + 5,$$
find the composite functions $f \circ g$ and $g \circ f$.

Solution We have
$$(f \circ g)(x) = f(g(x)) = \sqrt{2x^3+5},$$
$$(g \circ f)(x) = g(f(x)) = 2x^{3/2}+5.$$

It is possible to take the composition of three or more functions. For example, the composite function $f \circ g \circ h$ is found by first applying h, then g, and then f as follows:
$$(f \circ g \circ h)(x) = f(g(h(x))).$$

Example 2 Find $f \circ g \circ h$ if
$$f(x) = \ln x, \quad g(x) = 4 - x^2,$$
$$h(x) = \cos x.$$

Solution
$$(f \circ g \circ h)(x) = f(g(h(x)))$$
$$= f(4 - \cos^2 x)$$
$$= \ln(4 - \cos^2 x).$$

3. The Transformations of Functions

By applying certain transformations to the graph of a given function, we can obtain the graphs of certain related functions. This will give us the ability to sketch the graphs of many functions quickly by hand. It will also enable us to write equations for given graphs.

Let's firstly consider **translations**.

Vertical and Horizontal Shifts Suppose that $c > 0$.

To obtain the graph of
$$y = f(x) + c,$$
shift the graph of $y = f(x)$ a distance c units upward.

To obtain the graph of
$$y = f(x) - c,$$
shift the graph of $y = f(x)$ a distance c units downward.

To obtain the graph of
$$y = f(x - c),$$
shift the graph of $y = f(x)$ a distance c units to the right.

To obtain the graph of
$$y = f(x + c),$$
shift the graph of $y = f(x)$ a distance c units to the left.

解 我们有
$$(f \circ g)(x) = f(g(x)) = \sqrt{2x^3+5},$$
$$(g \circ f)(x) = g(f(x)) = 2x^{3/2}+5.$$

三个以上函数也可以复合. 例如, 为了计算复合函数
$$(f \circ g \circ h)(x) = f(g(h(x))),$$
先计算 h, 然后计算 g, 最后 f.

例 2 如果
$$f(x) = \ln x, \quad g(x) = 4 - x^2,$$
$$h(x) = \cos x,$$
求 $f \circ g \circ h$.

解
$$(f \circ g \circ h)(x) = f(g(h(x)))$$
$$= f(4 - \cos^2 x)$$
$$= \ln(4 - \cos^2 x).$$

3. 函数的变换

通过对一个给定函数的图形进行某一变换, 我们可以得到某一相关的函数图像. 这给我们提供了快速画出大量函数图像的能力, 也使我们能够根据给出的图像写出函数表达式.

首先考虑**平移变换**.

垂直和水平移动 设 $c > 0$.

为了得到
$$y = f(x) + c$$
的图像, 只需向上移动 $y = f(x)$ 的图像 c 个单位长度;

为了得到
$$y = f(x) - c$$
的图像, 只需向下移动 $y = f(x)$ 的图像 c 个单位长度;

为了得到
$$y = f(x - c)$$
的图像, 只需向右移动 $y = f(x)$ 的图像 c 个单位长度;

为了得到
$$y = f(x + c)$$
的图像, 只需向左移动 $y = f(x)$ 的图像 c 个单位长度.

Now let's consider the **stretching** and **reflecting transformations**.

The Vertical and Horizontal Stretching and the Reflecting

Suppose that $c>1$.

To obtain the graph of
$$y = cf(x),$$
stretch the graph of $y=f(x)$ vertically by a factor of c.

To obtain the graph of
$$y = (1/c)f(x),$$
compress the graph of $y=f(x)$ vertically by a factor of $1/c$.

To obtain the graph of
$$y = f(cx),$$
compress the graph of $y=f(x)$ horizontally by a factor of $1/c$.

To obtain the graph of
$$y = f(x/c),$$
stretch the graph of $y=f(x)$ horizontally by a factor of c.

To obtain the graph of
$$y = -f(x),$$
reflect the graph of $y=f(x)$ about the x-axis.

To obtain the graph of
$$y = f(-x),$$
reflect the graph of $y=f(x)$ about the y-axis.

现在考虑伸长与反射变换.

垂直和水平伸长与反射

设 $c>1$.

为了得到
$$y=cf(x)$$
的图像,只需垂直伸长 $y=f(x)$ 的图像到原来的 c 倍;

为了得到
$$y=(1/c)f(x)$$
的图像,只需垂直缩小 $y=f(x)$ 的图像到原来的 $1/c$ 倍;

为了得到
$$y=f(cx)$$
的图像,只需水平缩小 $y=f(x)$ 的图像到原来的 $1/c$ 倍;

为了得到
$$y=f(x/c)$$
的图像,只需水平伸长 $y=f(x)$ 的图像到原来的 c 倍;

为了得到
$$y=-f(x)$$
的图像,只需关于 x 轴反射 $y=f(x)$ 的图像;

为了得到
$$y=f(-x)$$
的图像,只需关于 y 轴反射 $y=f(x)$ 的图像.

1.4 Elementary Functions
1.4 初等函数

1. Basic Elementary Functions

(1) Constant functions:
$$y=C \quad (C \text{ is a constant}).$$

(2) Power functions:
$$y=x^a \quad (a \in \mathbf{R}).$$

(3) Exponential functions:
$$y=a^x \quad (a>0, a\neq 1).$$

(4) Logarithmic functions:
$$y=\log_a x \quad (a>0, a\neq 1).$$

1. 基本初等函数

(1) 常函数:
$$y=C \quad (C \text{ 是一个常数}).$$

(2) 幂函数:
$$y=x^a \quad (a \in \mathbf{R}).$$

(3) 指数函数:
$$y=a^x \quad (a>0, a\neq 1).$$

(4) 对数函数:
$$y=\log_a x \quad (a>0, a\neq 1).$$

(5) Trigonometric functions:
$$y=\sin x, \quad y=\cos x,$$
$$y=\tan x, \quad y=\cot x,$$
$$y=\sec x, \quad y=\csc x.$$

Here
$$\sec x=\frac{1}{\cos x}, \quad \csc x=\frac{1}{\sin x}.$$

The graphs of the functions $y=\sec x$ and $y=\csc x$ are shown in Figure 1.12 and Figure 1.13.

(5) 三角函数：
$$y=\sin x, \quad y=\cos x,$$
$$y=\tan x, \quad y=\cot x,$$
$$y=\sec x, \quad y=\csc x.$$

这里
$$\sec x=\frac{1}{\cos x}, \quad \csc x=\frac{1}{\sin x}.$$

函数 $y=\sec x$ 和 $y=\csc x$ 的图像分别如图 1.12 和图 1.13 所示.

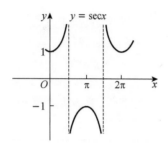

Figure 1.12

图 1.12

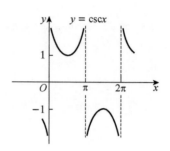

Figure 1.13

图 1.13

(6) Inverse trigonometric functions:

① $y=\arcsin x$ (Figure 1.14).

Domain: $[-1,1]$, range: $[-\pi/2, \pi/2]$.

② $y=\arccos x$ (Figure 1.15).

Domain: $[-1,1]$, range: $[0,\pi]$.

(6) 反三角函数：

① $y=\arcsin x$ （图 1.14）.

定义域：$[-1,1]$，值域：$[-\pi/2, \pi/2]$.

② $y=\arccos x$ （图 1.15）.

定义域：$[-1,1]$，值域：$[0,\pi]$.

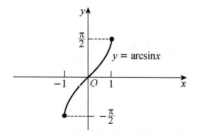

Figure 1.14

图 1.14

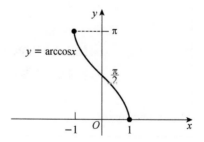

Figure 1.15

图 1.15

③ $y=\arctan x$ (Figure 1.16).

Domain: $(-\infty,+\infty)$, range: $(-\pi/2, \pi/2)$.

④ $y=\mathrm{arccot}\, x$ (Figure 1.17).

Domain: $(-\infty,+\infty)$, range: $(0,\pi)$.

③ $y=\arctan x$ （图 1.16）.

定义域：$(-\infty,+\infty)$，值域：$(-\pi/2, \pi/2)$.

④ $y=\mathrm{arccot}\, x$ （图 1.17）.

定义域：$(-\infty,+\infty)$，值域：$(0,\pi)$.

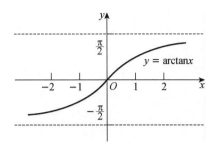

Figure 1.16
图 1.16

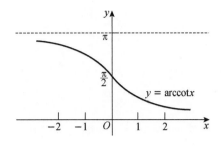

Figure 1.17
图 1.17

2. Elementary Functions

A function which is composed by basic elementary functions through finite arithmetic and compositions, and can be expressed only by one formula is called an **elementary function**.

For instance, the functions $y=\sqrt{1-x^2}, y=\sin^2 x, y=\sqrt{\cot\frac{x}{2}}$ are elementary functions, the function

$$f(x)=\begin{cases} x-1, & x<0, \\ 0, & x=0, \\ x+1, & x>0 \end{cases}$$

is not elementary function because it is not expressed by one formula.

2. 初等函数

由基本初等函数经过有限次的四则运算和有限次的函数复合步骤所构成并可用一个式子表示的函数，称为**初等函数**.

例如，函数 $y=\sqrt{1-x^2}$，$y=\sin^2 x$，$y=\sqrt{\cot\frac{x}{2}}$ 都是初等函数，函数

$$f(x)=\begin{cases} x-1, & x<0, \\ 0, & x=0, \\ x+1, & x>0 \end{cases}$$

不是初等函数，因为它不能由一个式子表示.

Exercises 1
习题 1

1. Find the domain of the following functions:
   (1) $f(x)=\dfrac{3}{5x^2+2x}$;
   (2) $f(x)=\sqrt{9-x^2}$;
   (3) $f(x)=\lg(4x-3)$;
   (4) $f(x)=\arcsin(x-3)$;
   (5) $f(x)=\dfrac{1}{x}-\sqrt{1-x^2}$;
   (6) $f(x)=\dfrac{x}{x^2-3x+2}$.

2. Determine whether the curve is the graph of a function of x. If it is, state the domain and range of the function.

1. 求出下列函数的定义域：
   (1) $f(x)=\dfrac{3}{5x^2+2x}$;
   (2) $f(x)=\sqrt{9-x^2}$;
   (3) $f(x)=\lg(4x-3)$;
   (4) $f(x)=\arcsin(x-3)$;
   (5) $f(x)=\dfrac{1}{x}-\sqrt{1-x^2}$;
   (6) $f(x)=\dfrac{x}{x^2-3x+2}$.

2. 判断下列曲线是否是函数的曲线. 如果是，请说出函数的定义域和值域.

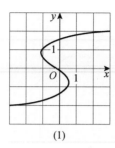
(1)

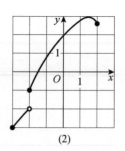
(2)

3. Determine whether each of the following functions is even, odd, or neither even nor odd.

(1) $y = 2x^3 - 8\sin x$;

(2) $y = a^x + a^{-x}$ $(a>0)$;

(3) $y = \dfrac{1-x^2}{1+x^2}$;

(4) $y = x(x-1)(x+1)$;

(5) $y = 2 + \cos x$;

(6) $y = \ln(x + \sqrt{x^2+1})$.

4. Find the functions

(1) $f \circ g$, (2) $g \circ f$, (3) $f \circ f$ and (4) $g \circ g$, where $f(x) = x^2 - 1$, $g(x) = 2x + 1$.

5. Given $F(x)$, find functions f, g and h such that $F = f \circ g \circ h$:

(1) $F(x) = \sqrt{\lg \sqrt{x}}$;

(2) $F(x) = \ln(\arccos x^3)$;

(3) $F(x) = \arctan^2(1+x)$;

(4) $F(x) = e^{\sqrt{1+x}}$.

6. The functions defined by the formulas

$$f(x) = \sqrt{x} \quad \text{and} \quad g(x) = \sqrt{1-x}.$$

Find $f+g, f-g, fg$ and f/g and state their domains.

3. 判断下面函数哪个是奇函数,哪个是偶函数,哪个是非奇非偶函数?

(1) $y = 2x^3 - 8\sin x$;

(2) $y = a^x + a^{-x}$ $(a>0)$;

(3) $y = \dfrac{1-x^2}{1+x^2}$;

(4) $y = x(x-1)(x+1)$;

(5) $y = 2 + \cos x$;

(6) $y = \ln(x + \sqrt{x^2+1})$.

4. 计算函数

(1) $f \circ g$; (2) $g \circ f$; (3) $f \circ f$; (4) $g \circ g$, 其中 $f(x) = x^2 - 1, g(x) = 2x + 1$.

5. 给定 $F(x)$,求出函数 f, g 和 h 使得 $F = f \circ g \circ h$:

(1) $F(x) = \sqrt{\lg \sqrt{x}}$;

(2) $F(x) = \ln(\arccos x^3)$;

(3) $F(x) = \arctan^2(1+x)$;

(4) $F(x) = e^{\sqrt{1+x}}$.

6. 定义函数 $f(x) = \sqrt{x}, g(x) = \sqrt{1-x}$,求 $f+g, f-g, fg$ 和 f/g,并说出它们的定义域.

Chapter 2   Limits
第 2 章   极限

2.1   The Limit of a Sequence
2.1   数列的极限

1. The Definition of the Convergent Sequence

Consider the following sequences:

(1) $\dfrac{1}{2}, \dfrac{2}{3}, \dfrac{3}{4}, \cdots, \dfrac{n}{n+1}, \cdots$;

(2) $\dfrac{1}{2}, \dfrac{1}{4}, \dfrac{1}{8}, \cdots, \dfrac{1}{2^n}, \cdots$;

(3) $1, -1, 1, -1, \cdots, (-1)^{n+1}, \cdots$;

(4) $2, \dfrac{1}{2}, \dfrac{4}{3}, \cdots, \dfrac{n+(-1)^{n-1}}{n}, \cdots$.

From the four sequences, we can see that the changing trend of the sequences are different as $n \to \infty$. In the sequence (1), $\dfrac{n}{n+1}$ close to 1 as $n \to \infty$; In the sequence (2), $\dfrac{1}{2^n}$ close to 0 as $n \to \infty$; In the sequence (3), $(-1)^{n+1}$ jumping between -1 and $+1$, while in the sequence (4), $\dfrac{n+(-1)^{n-1}}{n}$ jumping from the either side of 1, but still approach to 1 as n become sufficiently large.

In general, except from the sequence (3), the other three sequences have a determined change trend, that is when $n \to \infty$, the sequence approach to a determined constant L. We say the constant is the limit of the sequence.

Definition 2.1   A sequence $\{a_n\}$ has the limit L and we write
$$\lim_{n \to \infty} a_n = L \quad \text{or} \quad a_n \to L \ (n \to \infty)$$
if we can make the terms a_n as close to L as we like by taking n sufficiently large. If $\lim\limits_{n \to \infty} a_n$ exists, we say the sequence is **convergent**. Otherwise, we say the sequence is **divergent**.

1. 收敛数列的定义

考虑下面的数列:

(1) $\dfrac{1}{2}, \dfrac{2}{3}, \dfrac{3}{4}, \cdots, \dfrac{n}{n+1}, \cdots$;

(2) $\dfrac{1}{2}, \dfrac{1}{4}, \dfrac{1}{8}, \cdots, \dfrac{1}{2^n}, \cdots$;

(3) $1, -1, 1, -1, \cdots, (-1)^{n+1}, \cdots$;

(4) $2, \dfrac{1}{2}, \dfrac{4}{3}, \cdots, \dfrac{n+(-1)^{n-1}}{n}, \cdots$.

从这四个数列可以看出,当 $n \to \infty$ 时,它们的变化趋势是不相同的. 在数列(1)中,当 $n \to \infty$ 时,$\dfrac{n}{n+1}$ 趋于 1;在数列(2)中,当 $n \to \infty$ 时,$\dfrac{1}{2^n}$ 趋于 0;在数列(3)中,$(-1)^{n+1}$ 在 -1 和 $+1$ 间跳动;而在数列(4)中,当 n 足够大时,$\dfrac{n+(-1)^{n-1}}{n}$ 在 1 左、右两端来回跳动,但仍趋于 1.

总的来看,除数列(3)之外,其他三个数列在 $n \to \infty$ 时,都有一个确定的变化趋势,那就是在 $n \to \infty$ 时,数列都会趋向于一个确定的常数 L. 我们称这个常数为数列的极限.

定义 2.1   如果当 n 充分大时,a_n 无限接近常数 L,则称数列 $\{a_n\}$ 以 L 为**极限**,并记作
$$\lim_{n \to \infty} a_n = L \quad \text{或} \quad a_n \to L \ (n \to \infty).$$
如果数列 $\{a_n\}$ 的极限存在,则称该数列是**收敛**的;否则,称该数列是**发散**的.

If $\lim\limits_{n\to\infty}a_n=\infty$, then the sequence $\{a_n\}$ is divergent but in a special way. We say that $\{a_n\}$ diverges to ∞.

Definition 2.1 is a descriptive definition, a more precise mathematics definition is as follows:

Definition 2.2 A sequence $\{a_n\}$ has the **limit** L and we write
$$\lim_{n\to\infty}a_n=L \quad \text{or} \quad a_n\to L\ (n\to\infty)$$
if for every $\varepsilon>0$ there is a corresponding positive integer N such that if $n>N$ then
$$|a_n-L|<\varepsilon.$$

The geometric meaning of Definition 2.2 is shown in Figure 2.1, in which the terms $a_1, a_2, \cdots$ are plotted on a number line. No matter how small an interval $(L-\varepsilon, L+\varepsilon)$ is chosen, there exists an N such that all terms of the sequence from a_{N+1} onward must lie in this interval.

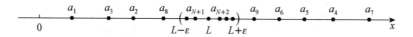

Figure 2.1

Example 1 Prove $\lim\limits_{n\to\infty}\dfrac{n+(-1)^{n-1}}{n}=1$.

Analysis Let ε be any positive number. We compute
$$|x_n-1|=\left|\dfrac{n+(-1)^{n-1}}{n}-1\right|=\dfrac{1}{n}.$$
$|x_n-1|<\varepsilon$ as long as $\dfrac{1}{n}<\varepsilon$, that is $n>\dfrac{1}{\varepsilon}$.

Proof For every $\varepsilon>0$, choose $N=\left[\dfrac{1}{\varepsilon}\right]+1\in\mathbf{Z}_+$, if $n>N$, we have
$$|x_n-1|=\left|\dfrac{n+(-1)^{n-1}}{n}-1\right|=\dfrac{1}{n}<\varepsilon,$$
thus
$$\lim_{n\to\infty}\dfrac{n+(-1)^{n-1}}{n}=1.$$

2. The Properties of a Convergent Sequence

(1) If a sequence $\{a_n\}$ is convergent, then the sequence $\{a_n\}$ is bounded.

(2) If a sequence $\{a_n\}$ is convergent, then the limit is unique.

For example, $\lim\limits_{n\to\infty}\dfrac{n}{n+1}=1$, there exists a positive number $M=2$, such that $\left|\dfrac{n}{n+1}\right|\leqslant 2$ for any positive number $n\in \mathbf{Z}_+$. Another example $\{(-1)^{n+1}\}$ is a divergent sequence, since if we say the sequence is convergent, then it is contradictory against the uniqueness of the limit.

2. 收敛数列的性质

（1）如果数列$\{a_n\}$收敛，那么数列$\{a_n\}$一定有界.

（2）如果数列$\{a_n\}$收敛，那么极限一定唯一.

例如，$\lim\limits_{n\to\infty}\dfrac{n}{n+1}=1$，存在一个正数 $M=2$，对于任意的正数 $n\in\mathbf{Z}_+$，使得 $\left|\dfrac{n}{n+1}\right|\leqslant 2$. 又如数列 $\{(-1)^{n+1}\}$ 是一个发散数列，因为如果我们说这个数列是收敛的，则与极限的唯一性矛盾.

2.2 The Limit of a Function
2.2 函数的极限

Having studied in the preceding section the limit of a sequence, we now turn our attention to limits in general functions.

上一节我们学习了数列的极限，这一节我们将注意力转向函数的极限.

1. The Limit of a Function as $x\to x_0$

Let us investigate the behavior of the function $f(x)$ defined by $f(x)=x^2-x+2$ for values of x near 2. Table 2.1 gives the values of $f(x)$ for values of x close to 2, but not equal to 2.

1. 函数在 $x\to x_0$ 时的极限

我们考查函数 $f(x)=x^2-x+2$ 在 $x=2$ 附近的行为（性态）. 表 2.1 给出了当 x 接近于 2 但是不等于 2 时函数值 $f(x)$ 的变化情况.

Table 2.1
表 2.1

x	$f(x)$	x	$f(x)$
1.0	2.000 000	3.0	8.000 000
1.5	2.750 000	2.5	5.750 000
1.8	3.440 000	2.2	4.640 000
1.9	3.710 000	2.1	4.310 000
1.95	3.852 500	2.05	4.152 500
1.99	3.970 100	2.01	4.030 100
1.995	3.985 025	2.005	4.015 025
1.999	3.997 001	2.001	4.003 001

From Table 2.1 and graph of f shown in Figure 2.2, we see that when x is close to 2, $f(x)$ is close to 4. We express this by saying "the limit equals to 4 as x approaches 2".

从表 2.1 和图 2.2 中我们看到,当 x 无限接近 2 时,$f(x)$ 无限接近 4. 我们说在 x 无限接近 2 时,函数 $f(x)=x^2-x+2$ 的极限等于 4.

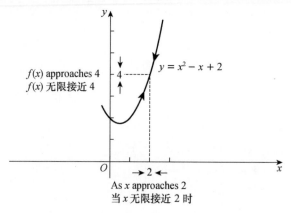

Figure 2.2

图 2.2

Definition 2.3 Suppose that $f(x)$ is defined on an open interval containing x_0, except possibly at x_0 itself. If $f(x)$ is arbitrarily close to L for all x sufficiently close to x_0. We say that $f(x)$ has the **limit** L as x approaches x_0, and write
$$\lim_{x \to x_0} f(x) = L.$$

An alternative notation for $\lim_{x \to x_0} f(x) = L$ is
$$f(x) \to L \quad (x \to x_0),$$
which is usually read "$f(x)$ approaches L as x approaches x_0".

A precise definition of a limit is as follows:

Definition 2.4 Let $f(x)$ be defined on an open interval containing x_0, except possibly at x_0 itself. We say that the **limit** of $f(x)$ as x approaches x_0 is the number L and write
$$\lim_{x \to x_0} f(x) = L$$
if for every $\varepsilon > 0$ there exists a corresponding number $\delta > 0$ such that for all x,
$$0 < |x - x_0| < \delta \Rightarrow |f(x) - L| < \varepsilon.$$

Notice that the phrase $0 < |x - x_0|$ in the definition of limit denote $x \neq x_0$. This means that the limit of a function $f(x)$ at a point x_0 has nothing to do with the value of

定义 2.3 设 $f(x)$ 除了可能在点 x_0 没有定义外,在包含 x_0 的一个开区间上均有定义. 如果对充分靠近 x_0 的 x,$f(x)$ 能任意靠近 L,那么称当 x 趋于 x_0 时 $f(x)$ 以 L 为**极限**,并记作
$$\lim_{x \to x_0} f(x) = L.$$
$\lim_{x \to x_0} f(x) = L$ 的另一种记法是
$$f(x) \to L \quad (x \to x_0),$$
读作"当 x 趋于 x_0 时,$f(x)$ 趋于 L".

极限的精确定义如下:

定义 2.4 设函数 $f(x)$ 定义在包含 x_0 的一个开区间上,可能在点 x_0 处没有定义. 如果对于任意的 $\varepsilon > 0$,都存在相应的数 $\delta > 0$,使得对所有满足 $0 < |x - x_0| < \delta$ 的 x,有
$$|f(x) - L| < \varepsilon,$$
则称当 x 趋于 x_0 时 $f(x)$ 以 L 为极限,并记作
$$\lim_{n \to \infty} f(x) = L.$$

注意到,定义中 $0 < |x - x_0|$ 表示 $x \neq x_0$. 这意味着函数在点 x_0 处的极限与该函数在点 x_0 处的值无关. 事实上,函数 $f(x)$ 在

the function at x_0. In fact, $f(x)$ need not even be defined when $x=x_0$.

The pictures in Figure 2.3 may help you absorb this definition.

$x=x_0$ 时可以没有定义.

图 2.3 可以帮助你理解定义 2.4.

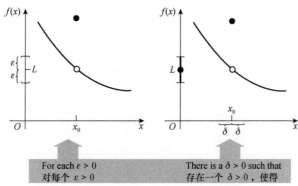

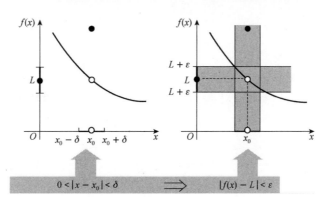

Figure 2.3

图 2.3

Example 1 Prove $\lim\limits_{x \to x_0} C = C$ (C is a constant).

Proof Here $|f(x)-L| = |C-C| = 0$. For any $\varepsilon > 0$, we can take any $\delta > 0$ such that
$$|f(x)-L| = |C-C| = 0 < \varepsilon$$
when $\quad 0 < |x-x_0| < \delta.$
Thus $\quad \lim\limits_{x \to x_0} C = C.$

Example 2 Prove $\lim\limits_{x \to x_0} x = x_0$.

Analysis $|f(x)-L| = |x-x_0|$. For every $\varepsilon > 0$, to make $|f(x)-L| < \varepsilon$, so long as $|x-x_0| < \varepsilon$.

Prove For every $\varepsilon > 0$, take $\delta = \varepsilon$, such that
$$|f(x)-L| = |x-x_0| < \varepsilon \quad \text{(Figure 2.4)}$$
when $\quad 0 < |x-x_0| < \delta.$

例 1 证明：$\lim\limits_{x \to x_0} C = C$（$C$ 是常数）.

证明 这里 $|f(x)-L| = |C-C| = 0$. 对于任意的 $\varepsilon > 0$，可任取 $\delta > 0$，当 $0 < |x-x_0| < \delta$ 时，有
$$|f(x)-L| = |C-C| = 0 < \varepsilon.$$
所以 $\quad \lim\limits_{x \to x_0} C = C.$

例 2 证明：$\lim\limits_{x \to x_0} x = x_0$.

分析 $|f(x)-L| = |x-x_0|$. 对任意的 $\varepsilon > 0$，要使 $|f(x)-L| < \varepsilon$，只要 $|x-x_0| < \varepsilon$.

证明 对任意的 $\varepsilon > 0$，取 $\delta = \varepsilon$，当 $0 < |x-x_0| < \delta$ 时，有
$$|f(x)-L| = |x-x_0| < \varepsilon \quad (\text{图 } 2.4).$$

Thus $$\lim_{x \to x_0} x = x_0.$$

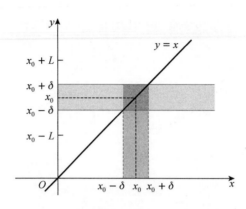

Figure 2.4

Example 3 Prove $\lim_{x \to 1}(2x-1)=1$.

Analysis $|f(x)-L|=|(2x-1)-1|=2|x-1|$. For every $\varepsilon > 0$, to make $|f(x)-L| < \varepsilon$, so long as $|x-1| < \frac{\varepsilon}{2}$.

Proof For every $\varepsilon > 0$, take $\delta = \frac{\varepsilon}{2}$, such that
$$|f(x)-L|=|(2x-1)-1|=2|x-1|<\varepsilon$$
when $0<|x-1|<\delta$.
Thus $\lim_{x \to 1}(2x-1)=1.$

2. One-sided Limits

To have a limit L as $x \to x_0$, a function $f(x)$ must be defined on both sides of x_0 and its values $f(x)$ must approach L approaches x_0 from either side. Because of this, ordinary limit are called two-sided.

Intuitively, if $f(x)$ is defined on an interval (c,b), where $c<b$, and approaches arbitrarily close to L as x approaches c from within that interval, then we say that $f(x)$ has a **right-hand limit** L as $x \to c$. We write
$$\lim_{x \to c^+} f(x) = L.$$
The symbol "$x \to c^+$" means x approaches c from the right.

Similarly, if $f(x)$ is defined on an interval (a, c), where $a < c$, and approaches arbitrarily close to M as x approaches c from within that interval, then we say that $f(x)$ has a **left-hand limit** M as $x \to c$. We write
$$\lim_{x \to c^-} f(x) = M.$$

The symbol "$x \to c^-$" means x approaches c from the left.

Left-hand limit and right-hand limit are collectively called **one-sided limits.**

These informal definitions of one-sided limits are illustrated in Figure 2.5.

类似地，设 $f(x)$ 定义在区间 (a, c) 上，其中 $a < c$，如果在该区间内，当 x 趋于 c 时，$f(x)$ 无限趋于 M，那么我们就说函数 $f(x)$ 在点 c 有**左侧极限** M，记作
$$\lim_{x \to c^-} f(x) = M.$$

符号"$x \to c^-$"表示 x 从左侧趋向于 c。

左侧极限和右侧极限统称为**单侧极限**。

这两个单侧极限的定义可用图 2.5 来说明。

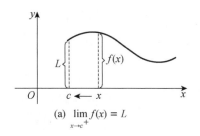

(a) $\lim_{x \to c^+} f(x) = L$ (b) $\lim_{x \to c^-} f(x) = M$

Figure 2.5
图 2.5

By comparing limit of function with the definitions of one-sided limits, we see that the following is true:

Theorem 2.1 A function $f(x)$ has a limit as $x \to c$ if and only if it has left-hand and right-hand limits there and these one-sided limits are equal:
$$\lim_{x \to c} f(x) = L \Leftrightarrow \lim_{x \to c^+} f(x) = \lim_{x \to c^-} f(x) = L.$$

Example 4 Prove that the limit of the function
$$f(x) = \begin{cases} x - 1, & x < 0, \\ 0, & x = 0, \\ x + 1, & x > 0 \end{cases}$$
as $x \to 0$ does not exist.

Proof The left hand limit of $f(x)$ as $x \to 0$ is
$$\lim_{x \to 0^-} f(x) = \lim_{x \to 0^-} (x - 1) = -1.$$
The right hand limit of $f(x)$ as $x \to 0$ is
$$\lim_{x \to 0^+} f(x) = \lim_{x \to 0^+} (x + 1) = 1.$$
Since $\lim_{x \to 0^-} f(x) \neq \lim_{x \to 0^+} f(x)$, using Theorem 2.1, the limit $\lim_{x \to 0} f(x)$ does not exist.

通过比较函数极限和单侧极限的定义，我们得出以下事实：

定理 2.1 函数 $f(x)$ 当 $x \to c$ 时有极限当且仅当函数的左、右极限都存在且相等，即
$$\lim_{x \to c} f(x) = L \Leftrightarrow \lim_{x \to c^+} f(x) = \lim_{x \to c^-} f(x) = L.$$

例 4 证明：函数
$$f(x) = \begin{cases} x - 1, & x < 0, \\ 0, & x = 0, \\ x + 1, & x > 0 \end{cases}$$
当 $x \to 0$ 时极限不存在。

证明 当 $x \to 0$ 时，$f(x)$ 的左极限为
$$\lim_{x \to 0^-} f(x) = \lim_{x \to 0^-} (x - 1) = -1.$$
当 $x \to 0$ 时，$f(x)$ 的右极限为
$$\lim_{x \to 0^+} f(x) = \lim_{x \to 0^+} (x + 1) = 1.$$
由于 $\lim_{x \to 0^-} f(x) \neq \lim_{x \to 0^+} f(x)$，根据定理 2.1，极限 $\lim_{x \to 0} f(x)$ 不存在。

3. The Limit of a Function as $x \to \infty$

If the values of $f(x)$ become closer and closer to L as x approaches infinity (write $x \to \infty$), we say that L is the limit of $f(x)$ as $x \to \infty$.

Definition 2.5 We say that $f(x)$ has the **limit** L as $x \to \infty$ and write
$$\lim_{x \to \infty} f(x) = L$$
if for every $\varepsilon > 0$, there exists a number $M > 0$ such that for all x,
$$|x| > M \Rightarrow |f(x) - L| < \varepsilon.$$

If $x > 0$ and x approaches infinity (write $x \to +\infty$), then we just replace $|x| > M$ by $x > M$ in Definition 2.5, we can get the definition of
$$\lim_{x \to +\infty} f(x) = L.$$

Similarly, if $x < 0$ and x approaches infinity (write $x \to -\infty$), $|x| > M$ replaced by $x < -M$ in Definition 2.5, we can get the definition of
$$\lim_{x \to -\infty} f(x) = L.$$

Looking at the graph of the function $f(x) = \dfrac{1}{x}$ (Figure 2.6), we observe that the x-axis is an asymptote of the curve on the right because $\lim\limits_{x \to +\infty} \dfrac{1}{x} = 0$. And on the left because $\lim\limits_{x \to -\infty} \dfrac{1}{x} = 0$, we say that the x-axis is a horizontal asymptote of the graph of $f(x) = \dfrac{1}{x}$. In fact, we have the following definition:

3. 函数在 $x \to \infty$ 时的极限

如果在 x 趋于无穷大(记作 $x \to \infty$)的过程中,对应的函数值 $f(x)$ 无限接近 L,那么称 L 为 $f(x)$ 当 $x \to \infty$ 时的极限.

定义 2.5 如果对于任意的 $\varepsilon > 0$,存在正数 $M > 0$,使得当 $|x| > M$ 时,都有
$$|f(x) - L| < \varepsilon,$$
则称函数 $f(x)$ 当 $x \to \infty$ 时以 L 为**极限**,记作
$$\lim_{x \to \infty} f(x) = L.$$

如果 $x > 0$ 且趋于无穷大(记作 $x \to +\infty$),我们只要把定义 2.5 中的 $|x| > M$ 改为 $x > M$,便得
$$\lim_{x \to +\infty} f(x) = L$$
的定义.

同样,如果 $x < 0$ 且 x 趋于无穷大(记作 $x \to -\infty$),我们只要把定义 2.5 中的 $|x| > M$ 改为 $x < -M$,便得
$$\lim_{x \to -\infty} f(x) = L$$
的定义.

观察函数 $f(x) = \dfrac{1}{x}$ 的图像(图 2.6),我们发现 x 轴是右侧曲线的一条渐近线,因为 $\lim\limits_{x \to \infty} \dfrac{1}{x} = 0$. 在左侧,由于 $\lim\limits_{x \to -\infty} \dfrac{1}{x} = 0$,我们说 x 轴是函数 $f(x) = \dfrac{1}{x}$ 的图像的一条水平渐近线.事实上,关于渐近线,我们有下面的定义:

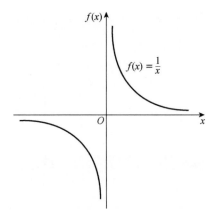

Figure 2.6
图 2.6

Definition 2.6 A line $y=b$ is called a **horizontal asymptote** of the graph of a function $y=f(x)$ if either
$$\lim_{x\to+\infty}f(x)=b \quad \text{or} \quad \lim_{x\to-\infty}f(x)=b.$$

4. Infinite Limits

Consider the function $f(x)=\dfrac{1}{x-2}$. As x gets close to 2 from the left, the function seems to decrease without bound. Similarly, as x approaches 2 from the right, the function seems to increase without bound. It therefore makes no sense to talk about $\lim\limits_{x\to 2}\dfrac{1}{x-2}$, but we think it is reasonable to write
$$\lim_{x\to 2^-}\frac{1}{x-2}=-\infty \quad \text{and} \quad \lim_{x\to 2^+}\frac{1}{x-2}=\infty.$$

Here is the precise definition:

Definition 2.7 We say that
$$\lim_{x\to c^+}f(x)=\infty$$
if for every positive number M, there exists a corresponding $\delta>0$ such that
$$0<x-c<\delta \Rightarrow f(x)>M.$$

Similarly, there are corresponding definitions of
$$\lim_{x\to c^+}f(x)=-\infty, \quad \lim_{x\to c^-}f(x)=\infty,$$
$$\lim_{x\to c^-}f(x)=-\infty, \quad \lim_{x\to c}f(x)=\infty,$$
$$\lim_{x\to\infty}f(x)=-\infty, \quad \lim_{x\to\infty}f(x)=\infty,$$
$$\lim_{x\to-\infty}f(x)=-\infty.$$

5. The Properties of Limits

(1) If $\lim\limits_{x\to x_0}f(x)$ exists, then the limit is unique.

(2) If $\lim\limits_{x\to x_0}f(x)=A$, then there exists two constants $M>0$ and $\delta>0$ such that
$$|f(x)|\leqslant M$$
as
$$0<|x-x_0|<\delta.$$

(3) If $\lim\limits_{x\to x_0}f(x)=A$ and $A>0$ (or $A<0$), then there exists a constant $\delta>0$ such that
$$f(x)>0 \quad (\text{or } f(x)<0)$$
as
$$0<|x-x_0|<\delta.$$

定义 2.6 如果函数 $y=f(x)$ 满足
$$\lim_{x\to+\infty}f(x)=b \quad \text{或} \quad \lim_{x\to-\infty}f(x)=b,$$
则称直线 $y=b$ 为函数 $y=f(x)$ 图像的**水平渐近线**.

4. 无穷极限

考虑函数 $f(x)=\dfrac{1}{x-2}$. 当 x 从左侧充分接近 2 时, 函数值无限减小. 类似地, 当 x 从右侧趋于 2 时, 函数值无限增大. 因此, 讨论 $\lim\limits_{x\to 2}\dfrac{1}{x-2}$ 毫无意义, 但是我们认为将这一现象记作
$$\lim_{x\to 2^-}\frac{1}{x-2}=-\infty \quad \text{和} \quad \lim_{x\to 2^+}\frac{1}{x-2}=\infty$$
是合理的. 这是精确的定义:

定义 2.7 如果对任何正实数 M, 存在相应的 $\delta>0$, 使得对一切满足 $0<x-c<\delta$ 的 x, 有
$$f(x)>M,$$
那么我们说
$$\lim_{x\to c^+}f(x)=\infty.$$

类似地, 有下面极限相应的定义:
$$\lim_{x\to c^+}f(x)=-\infty, \quad \lim_{x\to c^-}f(x)=\infty,$$
$$\lim_{x\to c^-}f(x)=-\infty, \quad \lim_{x\to c}f(x)=\infty,$$
$$\lim_{x\to\infty}f(x)=-\infty, \quad \lim_{x\to\infty}f(x)=\infty,$$
$$\lim_{x\to-\infty}f(x)=-\infty.$$

5. 极限的性质

(1) 如果 $\lim\limits_{x\to x_0}f(x)$ 存在, 那么此极限唯一.

(2) 如果 $\lim\limits_{x\to x_0}f(x)=A$, 那么存在常数 $M>0$ 和 $\delta>0$, 使得当 $0<|x-x_0|<\delta$ 时, 有
$$|f(x)|\leqslant M.$$

(3) 如果 $\lim\limits_{x\to x_0}f(x)=A$, 且 $A>0$ (或 $A<0$), 那么存在常数 $\delta>0$, 使得当 $0<|x-x_0|<\delta$ 时, 有
$$f(x)>0 \quad (\text{或 } f(x)<0).$$

2.3 Limit Laws
2.3 极限运算法则

Most readers have found that the "ε-δ" definition of the preceding section can prove the existence of values of limit, but we can not use the definition to compute the limits of functions. For the limit of easy functions, we can get the value of limit through observation, while for some complicate ones, observation method is both time consuming and difficult. This is why we will study the Limit Laws in this section. With them, we can handle most limit problems.

Limit Laws Suppose that c is a constant and the limits
$$\lim_{x \to a} f(x) \quad \text{and} \quad \lim_{x \to a} g(x)$$
exist, then
$$\lim_{x \to a} [f(x) + g(x)] = \lim_{x \to a} f(x) + \lim_{x \to a} g(x),$$
$$\lim_{x \to a} [f(x) - g(x)] = \lim_{x \to a} f(x) - \lim_{x \to a} g(x),$$
$$\lim_{x \to a} [cf(x)] = c \lim_{x \to a} f(x),$$
$$\lim_{x \to a} [f(x)g(x)] = \lim_{x \to a} f(x) \cdot \lim_{x \to a} g(x),$$
$$\lim_{x \to a} \frac{f(x)}{g(x)} = \frac{\lim_{x \to a} f(x)}{\lim_{x \to a} g(x)} \quad \text{if} \quad \lim_{x \to a} g(x) \neq 0.$$

These five laws can be stated verbally as follows:

The Sum Law: The limit of a sum is the sum of the limits;

The Difference Law: The limit of a difference is the difference of the limits;

The Constant Multiple Law: The limit of a constant times a function is the constant times the limit of the function;

The Product Law: The limit of a product is the product of the limits.

The Quotient Law: The limit of a quotient is the quotient of the limits (provided that the limit of the denominator is not 0).

大多数读者已经发现,上一节讲的极限的"ε-δ"定义可以证明极限值的存在性,不能用来求出函数的极限.对于简单函数的极限,我们可以通过观察得到,而对于一些形式复杂的函数的极限,利用观察法来求是很费时费力而且很困难的.这也是本节我们要学习极限运算法则的原因.利用极限运算法则,我们可以求出大多数函数的极限.

极限运算法测 假设 c 是一个常数,极限
$$\lim_{x \to a} f(x) \quad 和 \quad \lim_{x \to a} g(x)$$
存在,那么
$$\lim_{x \to a} [f(x) + g(x)] = \lim_{x \to a} f(x) + \lim_{x \to a} g(x),$$
$$\lim_{x \to a} [f(x) - g(x)] = \lim_{x \to a} f(x) - \lim_{x \to a} g(x),$$
$$\lim_{x \to a} [cf(x)] = c \lim_{x \to a} f(x),$$
$$\lim_{x \to a} [f(x)g(x)] = \lim_{x \to a} f(x) \cdot \lim_{x \to a} g(x),$$
$$\lim_{x \to a} \frac{f(x)}{g(x)} = \frac{\lim_{x \to a} f(x)}{\lim_{x \to a} g(x)}, \quad 当 \lim_{x \to a} g(x) \neq 0 \text{ 时}.$$

这5条法则可表述如下:

和法则:和的极限等于极限的和;

差法则:差的极限等于极限的差;

数乘法则:常数与函数乘积的极限等于这个常数与函数极限的乘积;

乘积法则:积的极限等于极限的积;

商法则:商的极限等于极限的商(假设分母的极限不等于0).

Remark 1 The Limit Laws not only tell us how to compute the limit of functions, but also tell us when the limit exist.

Remark 2 Although stated in terms of $x \to x_0$, the Limit Laws remain true for $x \to x_0^+$, $x \to x_0^-$, $x \to \infty$, $x \to +\infty$, $x \to -\infty$.

If we use the product law repeatedly with $g(x) = f(x)$, we obtain the following law:

The Power Law: $\lim\limits_{x \to a}[f(x)]^n = [\lim\limits_{x \to a} f(x)]^n$, where n is a positive integer.

In applying these six limit laws, we need to use two special limits:
$$\lim_{x \to a} C = C \ (C \text{ is a constant}),$$
$$\lim_{x \to a} x = a.$$

The following theorem give a additional property of limits:

Theorem 2.2 If $\varphi(x) \geqslant \psi(x)$, and
$$\lim_{x \to x_0} \varphi(x) = a, \quad \lim_{x \to x_0} \psi(x) = b,$$
then $a \geqslant b$.

Example 1 Find $\lim\limits_{x \to 1}(2x-1)$.

Solution
$$\begin{aligned}
\lim_{x \to 1}(2x-1) &= \lim_{x \to 1} 2x - \lim_{x \to 1} 1 \\
&= 2 \lim_{x \to 1} x - 1 \\
&= 2 \cdot 1 - 1 = 1.
\end{aligned}$$

Example 2 Find $\lim\limits_{x \to 2} \dfrac{x^3-1}{x^2-5x+3}$.

Solution
$$\begin{aligned}
\lim_{x \to 2} \frac{x^3-1}{x^2-5x+3} &= \frac{\lim\limits_{x \to 2}(x^3-1)}{\lim\limits_{x \to 2}(x^2-5x+3)} \\
&= \frac{\lim\limits_{x \to 2} x^3 - \lim\limits_{x \to 2} 1}{\lim\limits_{x \to 2} x^2 - 5 \lim\limits_{x \to 2} x + \lim\limits_{x \to 2} 3} \\
&= \frac{(\lim\limits_{x \to 2} x)^3 - 1}{(\lim\limits_{x \to 2} x)^2 - 5 \cdot 2 + 3} \\
&= \frac{2^3-1}{2^2-10+3} = -\frac{7}{3}.
\end{aligned}$$

From above two examples, we can see that in order to get the limit of polynomial and rational functions as $x \to x_0$, we just substitute x by x_0. But for rational function, if the value of denominator at x_0 is zero, then there is no significance.

注1 极限运算法则不仅仅告诉我们极限如何计算,而且告诉我们极限什么时候存在.

注2 虽然在法则中仅仅标明 $x \to x_0$,但是极限的运算法则对于 $x \to x_0^+$, $x \to x_0^-$, $x \to \infty$, $x \to +\infty$, $x \to -\infty$ 的情形都是成立的.

在乘积法则中,如果我们令 $g(x) = f(x)$,并反复利用乘积法则,则得到下面的法则:

乘幂法则: $\lim\limits_{x \to a}[f(x)]^n = [\lim\limits_{x \to a} f(x)]^n$,其中 n 是正整数.

在上面6条极限法则的运用中,我们需要用到两个特殊的极限:
$$\lim_{x \to a} C = C \ (C \text{ 是常数}),$$
$$\lim_{x \to a} x = a.$$

下面的定理给出了极限的另外一条性质:

定理2.2 如果 $\varphi(x) \geqslant \psi(x)$,并且
$$\lim_{x \to x_0} \varphi(x) = a, \quad \lim_{x \to x_0} \psi(x) = b,$$
那么 $a \geqslant b$.

例1 求 $\lim\limits_{x \to 1}(2x-1)$.

解
$$\begin{aligned}
\lim_{x \to 1}(2x-1) &= \lim_{x \to 1} 2x - \lim_{x \to 1} 1 \\
&= 2 \lim_{x \to 1} x - 1 \\
&= 2 \cdot 1 - 1 = 1.
\end{aligned}$$

例2 求 $\lim\limits_{x \to 2} \dfrac{x^3-1}{x^2-5x+3}$.

解
$$\begin{aligned}
\lim_{x \to 2} \frac{x^3-1}{x^2-5x+3} &= \frac{\lim\limits_{x \to 2}(x^3-1)}{\lim\limits_{x \to 2}(x^2-5x+3)} \\
&= \frac{\lim\limits_{x \to 2} x^3 - \lim\limits_{x \to 2} 1}{\lim\limits_{x \to 2} x^2 - 5 \lim\limits_{x \to 2} x + \lim\limits_{x \to 2} 3} \\
&= \frac{(\lim\limits_{x \to 2} x)^3 - 1}{(\lim\limits_{x \to 2} x)^2 - 5 \cdot 2 + 3} \\
&= \frac{2^3-1}{2^2-10+3} = -\frac{7}{3}.
\end{aligned}$$

从上面的两个例子可以看出,求多项式或有理分式函数当 $x \to x_0$ 时的极限时,只需要用 x_0 代替函数中的 x 就行了.但是,对于有理函数,代入 x_0 后如果分母等于零,则没有意义.

The Direct Substitution Property If $f(x)$ is a polynomial or a rational function and x_0 is in the domain of $f(x)$, then
$$\lim_{x\to x_0} f(x) = f(x_0).$$

In fact, if the polynomial is
$$P(x) = a_0 x^n + a_1 x^{n-1} + \cdots + a_{n-1} x + a_n,$$
then
$$\begin{aligned}\lim_{x\to x_0} P(x) &= \lim_{x\to x_0} a_0 x^n + \lim_{x\to x_0} a_1 x^{n-1} + \cdots \\ &\quad + \lim_{x\to x_0} a_{n-1} x + \lim_{x\to x_0} a_n \\ &= a_0 \lim_{x\to x_0} x^n + a_1 \lim_{x\to x_0} x^{n-1} + \cdots \\ &\quad + a_{n-1} \lim_{x\to x_0} x + \lim_{x\to x_0} a_n \\ &= a_0 (\lim_{x\to x_0} x)^n + a_1 (\lim_{x\to x_0} x)^{n-1} + \cdots + a_n \\ &= a_0 x_0^n + a_1 x_0^{n-1} + \cdots + a_{n-1} x_0 + a_n \\ &= P(x_0).\end{aligned}$$

If the rational function is
$$F(x) = \frac{P(x)}{Q(x)},$$
where $P(x)$ and $Q(x)$ are both polynomials, then
$$\lim_{x\to x_0} P(x) = P(x_0), \quad \lim_{x\to x_0} Q(x) = Q(x_0),$$
when $Q(x_0) \neq 0$,
$$\lim_{x\to x_0} \frac{P(x)}{Q(x)} = \frac{\lim_{x\to x_0} P(x)}{\lim_{x\to x_0} Q(x)} = \frac{P(x_0)}{Q(x_0)} = F(x_0).$$

However, not all limits can be evaluated direct substitute.

Example 3 Find $\lim\limits_{x\to 3} \dfrac{x-3}{x^2-9}$.

Solution Let $f(x) = \dfrac{x-3}{x^2-9}$. We can not find the limit by substituting $x=3$ because $f(3)$ is not defined. We also can not apply the Quotient Law because the limit of the denominator is 0. Instead, we could factor denominator using squared difference:
$$f(x) = \frac{x-3}{(x-3)(x+3)}.$$

The numerator and denominator have a common factor:

直接代入性质 如果 $f(x)$ 是一个多项式或有理函数，x_0 在 $f(x)$ 的定义域内，那么
$$\lim_{x\to x_0} f(x) = f(x_0).$$

事实上，如果多项式为
$$P(x) = a_0 x^n + a_1 x^{n-1} + \cdots + a_{n-1} x + a_n,$$
那么
$$\begin{aligned}\lim_{x\to x_0} P(x) &= \lim_{x\to x_0} a_0 x^n + \lim_{x\to x_0} a_1 x^{n-1} + \cdots \\ &\quad + \lim_{x\to x_0} a_{n-1} x + \lim_{x\to x_0} a_n \\ &= a_0 \lim_{x\to x_0} x^n + a_1 \lim_{x\to x_0} x^{n-1} + \cdots \\ &\quad + a_{n-1} \lim_{x\to x_0} x + \lim_{x\to x_0} a_n \\ &= a_0 (\lim_{x\to x_0} x)^n + a_1 (\lim_{x\to x_0} x)^{n-1} + \cdots + a_n \\ &= a_0 x_0^n + a_1 x_0^{n-1} + \cdots + a_{n-1} x_0 + a_n \\ &= P(x_0).\end{aligned}$$

如果有理函数为
$$F(x) = \frac{P(x)}{Q(x)},$$
其中 $P(x)$ 和 $Q(x)$ 都是多项式，那么有
$$\lim_{x\to x_0} P(x) = P(x_0), \quad \lim_{x\to x_0} Q(x) = Q(x_0),$$
当 $Q(x_0) \neq 0$ 时，有
$$\lim_{x\to x_0} \frac{P(x)}{Q(x)} = \frac{\lim_{x\to x_0} P(x)}{\lim_{x\to x_0} Q(x)} = \frac{P(x_0)}{Q(x_0)} = F(x_0).$$

然而，并不是所有的有理函数的极限都可以用直接替代方法.

例3 求 $\lim\limits_{x\to 3} \dfrac{x-3}{x^2-9}$.

解 令 $f(x) = \dfrac{x-3}{x^2-9}$. 我们不能直接代入 $x=3$，因为 $f(3)$ 是没有定义的. 我们也不能利用商法则，因为分母的极限为 0. 反而，我们可以利用平方差公式将函数的分母进行因式分解：
$$f(x) = \frac{x-3}{(x-3)(x+3)}.$$

分子和分母有公因子 $x-3$. 当 $x\to 3$ 时，$x\neq 3$,

$x-3$. When taking the limit as x approaches 3, we have $x \neq 3$ and so $x-3 \neq 0$. Therefore we can cancel the common factor and compute the limit as follow:

$$\lim_{x \to 3} \frac{x-3}{x^2-9} = \lim_{x \to 3} \frac{x-3}{(x-3)(x+3)}$$
$$= \lim_{x \to 3} \frac{1}{x+3}$$
$$= \frac{\lim_{x \to 3} 1}{\lim_{x \to 3}(x+3)} = \frac{1}{6}.$$

Example 4 Find $\lim\limits_{x \to \infty} \dfrac{3x^3+4x^2+2}{7x^3+5x^2-3}$.

Solution $\lim\limits_{x \to \infty} \dfrac{3x^3+4x^2+2}{7x^3+5x^2-3} = \lim\limits_{x \to \infty} \dfrac{3+\dfrac{4}{x}+\dfrac{2}{x^3}}{7+\dfrac{5}{x}-\dfrac{3}{x^3}} = \dfrac{3}{7}$.

Example 5 Find $\lim\limits_{x \to \infty} \dfrac{3x^2-2x-1}{2x^3-x^2+5}$.

Solution $\lim\limits_{x \to \infty} \dfrac{3x^2-2x-1}{2x^3-x^2+5} = \lim\limits_{x \to \infty} \dfrac{\dfrac{3}{x}-\dfrac{2}{x^2}-\dfrac{1}{x^3}}{2-\dfrac{1}{x}+\dfrac{5}{x^3}}$

$$= \frac{0}{2} = 0.$$

Example 6 Find $\lim\limits_{x \to \infty} \dfrac{2x^3-x^2+5}{3x^2-2x-1}$.

Solution Since $\lim\limits_{x \to \infty} \dfrac{3x^2-2x-1}{2x^3-x^2+5} = 0$, thus

$$\lim_{x \to \infty} \frac{2x^3-x^2+5}{3x^2-2x-1} = \infty.$$

Example 4~6 are special cases of the following limits of rational functions:

$$\lim_{x \to \infty} \frac{a_0 x^n + a_1 x^{n-1} + \cdots + a_n}{b_0 x^m + b_1 x^{m-1} + \cdots + b_m} = ?$$

When $a_0 \neq 0, b_0 \neq 0, m, n$ are non-negative integers, we have the following results:

$$\lim_{x \to \infty} \frac{a_0 x^n + a_1 x^{n-1} + \cdots + a_n}{b_0 x^m + b_1 x^{m-1} + \cdots + b_m} = \begin{cases} 0, & n < m, \\ \dfrac{a_0}{b_0}, & n = m, \\ \infty, & n > m. \end{cases}$$

所以 $x-3 \neq 0$,因此可约去公因子,如下计算极限:

$$\lim_{x \to 3} \frac{x-3}{x^2-9} = \lim_{x \to 3} \frac{x-3}{(x-3)(x+3)}$$
$$= \lim_{x \to 3} \frac{1}{x+3}$$
$$= \frac{\lim_{x \to 3} 1}{\lim_{x \to 3}(x+3)} = \frac{1}{6}.$$

例 4 求 $\lim\limits_{x \to \infty} \dfrac{3x^3+4x^2+2}{7x^3+5x^2-3}$.

解 $\lim\limits_{x \to \infty} \dfrac{3x^3+4x^2+2}{7x^3+5x^2-3} = \lim\limits_{x \to \infty} \dfrac{3+\dfrac{4}{x}+\dfrac{2}{x^3}}{7+\dfrac{5}{x}-\dfrac{3}{x^3}} = \dfrac{3}{7}$.

例 5 求 $\lim\limits_{x \to \infty} \dfrac{3x^2-2x-1}{2x^3-x^2+5}$.

解 $\lim\limits_{x \to \infty} \dfrac{3x^2-2x-1}{2x^3-x^2+5} = \lim\limits_{x \to \infty} \dfrac{\dfrac{3}{x}-\dfrac{2}{x^2}-\dfrac{1}{x^3}}{2-\dfrac{1}{x}+\dfrac{5}{x^3}}$

$$= \frac{0}{2} = 0.$$

例 6 求 $\lim\limits_{x \to \infty} \dfrac{2x^3-x^2+5}{3x^2-2x-1}$.

解 因为 $\lim\limits_{x \to \infty} \dfrac{3x^2-2x-1}{2x^3-x^2+5} = 0$,所以

$$\lim_{x \to \infty} \frac{2x^3-x^2+5}{3x^2-2x-1} = \infty.$$

例 4~例 6 是下列有理函数的极限的特例:

$$\lim_{x \to \infty} \frac{a_0 x^n + a_1 x^{n-1} + \cdots + a_n}{b_0 x^m + b_1 x^{m-1} + \cdots + b_m} = ?$$

当 $a_0 \neq 0, b_0 \neq 0, m, n$ 为非负整数时,有

$$\lim_{x \to \infty} \frac{a_0 x^n + a_1 x^{n-1} + \cdots + a_n}{b_0 x^m + b_1 x^{m-1} + \cdots + b_m} = \begin{cases} 0, & n < m, \\ \dfrac{a_0}{b_0}, & n = m, \\ \infty, & n > m. \end{cases}$$

2.4 Limit Existence Rules and Two Important Limits
2.4 极限存在准则和两个重要极限

In this section, we will study two principles for deciding the existence of limit, and with two principles, we discuss two important limits:

$$\lim_{\theta \to 0} \frac{\sin\theta}{\theta} = 1 \quad \text{and} \quad \lim_{x \to \infty} \left(1 + \frac{1}{x}\right)^x = e.$$

Principle I Suppose that the sequences $\{x_n\}, \{y_n\}$ and $\{z_n\}$ satisfy the following hypotheses:

(1) there exists $n_0 \in \mathbf{Z}_+$, such that
$$y_n \leqslant x_n \leqslant z_n \quad \text{as} \quad n > n_0;$$
(2) $\lim_{n \to \infty} y_n = a$, $\lim_{n \to \infty} z_n = a$,

then $\lim_{n \to \infty} x_n$ exists, and $\lim_{n \to \infty} x_n = a$.

We can generalize the result of Principle I for sequences to the limit of functions.

Principle II Suppose that functions $g(x)$, $f(x)$ and $h(x)$ satisfy the following hypotheses:

(1) for all x in some open interval containing x_0 (except maybe x_0),
$$g(x) \leqslant f(x) \leqslant h(x);$$
(2) $\lim_{x \to x_0} g(x) = \lim_{x \to x_0} h(x) = L,$

then
$$\lim_{x \to x_0} f(x) = L.$$

Principle II says that if $f(x)$ is squeezed between $g(x)$ and $h(x)$ in some open interval containing x_0, and if $g(x)$ and $h(x)$ have the same limit L at x_0, then $f(x)$ is forced to have the same limit L at x_0 (Figure 2.7). For $x \to x_0^+, x \to x_0^-, x \to +\infty, x \to -\infty$, there are corresponding limit existence rules.

Principle I and Principle II are called the **Squeeze Principle**, **Pinching Principle** or **Sandwich Principle**.

Example 1 Prove $\lim_{x \to 0} x^2 \sin \frac{1}{x} = 0$.

在这一节中,我们学习判定极限存在的两个准则,并利用这两个准则,讨论两个重要极限:

$$\lim_{\theta \to 0} \frac{\sin\theta}{\theta} = 1 \quad 和 \quad \lim_{x \to \infty} \left(1 + \frac{1}{x}\right)^x = e.$$

准则 I 假设数列 $\{x_n\}, \{y_n\}$ 和 $\{z_n\}$ 满足下列条件:

(1) 存在 $n_0 \in \mathbf{Z}_+$,使得当 $n > n_0$ 时,有
$$y_n \leqslant x_n \leqslant z_n;$$
(2) $\lim_{n \to \infty} y_n = a$, $\lim_{n \to \infty} z_n = a$,

则 $\lim_{n \to \infty} x_n$ 存在,且 $\lim_{n \to \infty} x_n = a$.

我们可以将对于数列的准则 I 的结果推广到函数的极限.

准则 II 假设函数 $g(x), f(x)$ 和 $h(x)$ 满足下列条件:

(1) 对包含 x_0 的某个开区间(x_0 可能除外)内的所有 x,成立
$$g(x) \leqslant f(x) \leqslant h(x);$$
(2) $\lim_{x \to x_0} g(x) = \lim_{x \to x_0} h(x) = L,$

则
$$\lim_{x \to x_0} f(x) = L.$$

准则 II 的意思是,在包含 x_0 的某个开区间内,如果 $f(x)$ 被夹在 $g(x)$ 和 $h(x)$ 之间,并且 $g(x)$ 和 $h(x)$ 在 x_0 处有相同的极限 L,那么 $f(x)$ 在 x_0 处也有相同的极限 L(图 2.7).对于 $x \to x_0^+, x \to x_0^-, x \to \infty, x \to +\infty, x \to -\infty$ 的情形,也有相应的极限存在准则.

准则 I 和准则 II 称为**夹逼准则**、**夹挤准则**或**迫近准则**.

例 1 证明: $\lim_{x \to 0} x^2 \sin \frac{1}{x} = 0$.

2.4 Limit Existence Rules and Two Important Limits
2.4 极限存在准则和两个重要极限

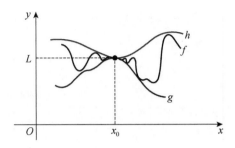

Figure 2.7

图 2.7

Proof Since
$$-1 \leqslant \sin \frac{1}{x} \leqslant 1,$$
we have
$$-x^2 \leqslant x^2 \sin \frac{1}{x} \leqslant x^2,$$
as illustrated by Figure 2.8. We know that
$$\lim_{x \to 0} x^2 = 0 \quad \text{and} \quad \lim_{x \to 0}(-x^2) = 0.$$
From the Squeeze Principle, we obtain
$$\lim_{x \to 0} x^2 \sin \frac{1}{x} = 0.$$

证明 由于
$$-1 \leqslant \sin \frac{1}{x} \leqslant 1,$$
如图 2.8 所示，我们有
$$-x^2 \leqslant x^2 \sin \frac{1}{x} \leqslant x^2.$$
易知
$$\lim_{x \to 0} x^2 = 0 \quad \text{和} \quad \lim_{x \to 0}(-x^2) = 0.$$
利用夹逼准则，我们得到
$$\lim_{x \to 0} x^2 \sin \frac{1}{x} = 0.$$

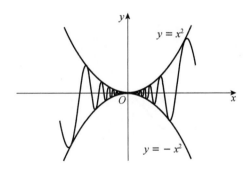

Figure 2.8

图 2.8

As application of the Squeeze Principle, we prove an important limit:
$$\lim_{\theta \to 0} \frac{\sin \theta}{\theta} = 1.$$

Proof In order to prove
$$\lim_{\theta \to 0} \frac{\sin \theta}{\theta} = 1,$$

作为夹逼准则的应用，我们来证明一个重要的极限
$$\lim_{\theta \to 0} \frac{\sin \theta}{\theta} = 1.$$

证明 为了证明
$$\lim_{\theta \to 0} \frac{\sin \theta}{\theta} = 1,$$

we are going to show that the right-hand and left-hand limits of $\dfrac{\sin\theta}{\theta}$ at 0 are both 1. To show that $\lim\limits_{\theta\to 0^+}\dfrac{\sin\theta}{\theta}=1$ we assume that θ less than $\dfrac{\pi}{2}$ (Figure 2.9). Notice that

$$\text{The area of } \triangle OAP < \text{The area of sector } OAP$$
$$< \text{The area of } \triangle OAT,$$

thus

$$\frac{1}{2}\sin\theta < \frac{1}{2}\theta < \frac{1}{2}\tan\theta.$$

This inequality goes the same way if we divide all three terms by the number $\dfrac{1}{2}\sin\theta$, which is positive since $0<\theta<\dfrac{\pi}{2}$:

$$1 < \frac{\theta}{\sin\theta} < \frac{1}{\cos\theta}.$$

Taking reciprocals reverses the inequalities:

$$\cos\theta < \frac{\sin\theta}{\theta} < 1. \qquad (2.1)$$

From $\lim\limits_{\theta\to 0^+}\dfrac{1}{\cos\theta}=1$ and the Squeeze Principle, we have

$$\lim_{\theta\to 0^+}\frac{\sin\theta}{\theta}=1.$$

我们证明 $\dfrac{\sin\theta}{\theta}$ 在 $x=0$ 处的左、右极限都等于 1. 为了证明 $\lim\limits_{\theta\to 0^+}\dfrac{\sin\theta}{\theta}=1$,我们假设 $0<\theta<\dfrac{\pi}{2}$(图 2.9). 注意到

$$\triangle OAP \text{ 的面积} < \text{扇形 } OAP \text{ 的面积}$$
$$< \triangle OAT \text{ 的面积},$$

因此

$$\frac{1}{2}\sin\theta < \frac{1}{2}\theta < \frac{1}{2}\tan\theta.$$

由于 $0<\theta<\dfrac{\pi}{2}$,因此 $\dfrac{1}{2}\sin\theta>0$. 对上述不等式的三项同时除以 $\dfrac{1}{2}\sin\theta$,不等式仍然成立:

$$1 < \frac{\theta}{\sin\theta} < \frac{1}{\cos\theta}.$$

不等式中各项取倒数,得到

$$\cos\theta < \frac{\sin\theta}{\theta} < 1. \qquad (2.1)$$

由 $\lim\limits_{\theta\to 0^+}\dfrac{1}{\cos\theta}=1$ 和夹逼准则,我们有

$$\lim_{\theta\to 0^+}\frac{\sin\theta}{\theta}=1.$$

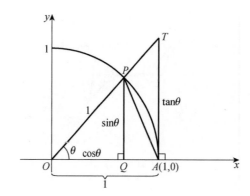

Figure 2.9
图 2.9

Next, we will prove

$$\lim_{\theta\to 0^-}\frac{\sin\theta}{\theta}=1.$$

We should have noticed that the inequality (2.1) holds when

下面证明

$$\lim_{\theta\to 0^-}\frac{\sin\theta}{\theta}=1.$$

我们应该注意到,由于 $\dfrac{\theta}{\sin\theta}$ 是偶函数,不等式

$-\frac{\pi}{2}<\theta<0$ since $\frac{\theta}{\sin\theta}$ is an even function. Thus
$$\lim_{\theta\to 0^-}\frac{\sin\theta}{\theta}=1.$$
Consequently, we have
$$\lim_{\theta\to 0}\frac{\sin\theta}{\theta}=1.$$

Example 2 Find $\lim\limits_{x\to 0}\dfrac{\tan x}{x}$.

Solution
$$\lim_{x\to 0}\frac{\tan x}{x}=\lim_{x\to 0}\frac{\sin x}{x}\cdot\frac{1}{\cos x}$$
$$=\lim_{x\to 0}\frac{\sin x}{x}\cdot\lim_{x\to 0}\frac{1}{\cos x}=1.$$

Example 3 Find $\lim\limits_{x\to 0}\dfrac{1-\cos x}{x^2}$.

Solution
$$\lim_{x\to 0}\frac{1-\cos x}{x^2}=\lim_{x\to 0}\frac{2\sin^2\frac{x}{2}}{x^2}$$
$$=\frac{1}{2}\lim_{x\to 0}\frac{\sin^2\frac{x}{2}}{\left(\frac{x}{2}\right)^2}$$
$$=\frac{1}{2}\lim_{x\to 0}\left(\frac{\sin\frac{x}{2}}{\frac{x}{2}}\right)^2$$
$$=\frac{1}{2}\cdot 1^2=\frac{1}{2}.$$

A sequence $\{a_n\}$ is called **increasing** if
$$a_1<a_2<\cdots<a_n<a_{n+1}<\cdots.$$
It is called **decreasing** if
$$a_1>a_2>\cdots>a_n>a_{n+1}>\cdots.$$

The increasing sequence and the decreasing sequence are collectively called the **monotonic sequences**.

Principle Ⅲ Every bounded, monotonic sequence is convergent.

We know that in Section 2.1, if a sequence $\{a_n\}$ is convergent, then the sequence is bounded. But the converse result may be not true. Principle Ⅲ tells us if the sequence is bounded, and monotonic, then the limit of the sequence exists, that is the sequence is convergent.

From Principle Ⅲ, we can prove another important limit:
$$\lim_{n\to\infty}\left(1+\frac{1}{n}\right)^n = e.$$
Here e is an irrational number:
$$e = 2.718\ 281\ 828\ 459\ 045\cdots.$$
We can also prove that
$$\lim_{x\to\infty}\left(1+\frac{1}{x}\right)^x = e,$$
$$\lim_{x\to 0}(1+x)^{\frac{1}{x}} = e.$$

Example 4 Find $\lim_{x\to\infty}\left(1-\frac{1}{x}\right)^x$.

Solution Let $t=-x$, then $t\to\infty$ as $x\to\infty$. Thus
$$\lim_{x\to\infty}\left(1-\frac{1}{x}\right)^x = \lim_{t\to\infty}\left(1+\frac{1}{t}\right)^{-t}$$
$$= \lim_{t\to\infty}\frac{1}{\left(1+\frac{1}{t}\right)^t} = \frac{1}{e}.$$

2.5 The Continuity of Functions

In Section 2.3, we noticed that the limit of a function as x approaches x_0 can often be found simply by calculating the value of the function at x_0. For instance, If $f(x)$ is a polynomial or a rational function and x_0 is in the domain of $f(x)$, then
$$\lim_{x\to x_0}f(x) = f(x_0).$$
The functions with this property are called continuous at the point x_0. Intuitively, any function $y=f(x)$ whose graph can be sketched over its domain in one continuous motion without lifting the pencil is an example of a continuous function.

In this section, we investigate more precisely what it means for a function to be continuous. We also study the properties of continuous functions, and see that many of the function types presented in Section 1.1 are continuous.

利用准则Ⅲ，我们可以证明另一个重要极限：
$$\lim_{n\to\infty}\left(1+\frac{1}{n}\right)^n = e.$$
这里 e 是无理数：
$$e = 2.718\ 281\ 828\ 459\ 045\cdots.$$
我们也可以证明：
$$\lim_{x\to\infty}\left(1+\frac{1}{x}\right)^x = e,$$
$$\lim_{x\to 0}(1+x)^{\frac{1}{x}} = e.$$

例4 计算 $\lim_{x\to\infty}\left(1-\frac{1}{x}\right)^x$.

解 令 $t=-x$，则当 $x\to\infty$ 时，$t\to\infty$. 因此
$$\lim_{x\to\infty}\left(1-\frac{1}{x}\right)^x = \lim_{t\to\infty}\left(1+\frac{1}{t}\right)^{-t}$$
$$= \lim_{t\to\infty}\frac{1}{\left(1+\frac{1}{t}\right)^t} = \frac{1}{e}.$$

2.5 函数的连续性

在 2.3 节中，我们注意到，函数当 x 趋于 x_0 时的极限值有时简单到可以直接通过计算函数在 x_0 处的函数值得到. 例如，如果 $f(x)$ 是一个多项式或有理函数，并且 x_0 在其定义域中，那么有
$$\lim_{x\to x_0}f(x) = f(x_0).$$
我们称具有这种性质的函数在点 x_0 处连续. 直观上来看，对于任何函数 $y=f(x)$，如果在其定义域上它的图像可以由铅笔在纸上连续运动而始终不离开纸画出来的话，这一函数就是连续函数的一个例子.

在本节中，我们将学习连续函数的更精确的定义. 同时，我们也学习连续函数的性质，并且会看到在 1.1 节中涉及的很多函数都是连续的.

2.5 The Continuity of Functions
2.5 函数的连续性

1. Continuity at a Point

To define continuity in a function's domain D, we need to define continuity at an interior point x_0 (that is, there exists $\delta > 0$, such that $(x_0-\delta, x_0+\delta) \subset D$).

Definition 2.8 We say that a function $f(x)$ is **continuous** at an interior point x_0 of its domain if
$$\lim_{x \to x_0} f(x) = f(x_0).$$

Example 1 Figure 2.10 shows the graph of a function $f(x)$. At which numbers is $f(x)$ discontinuous? Why?

1. 在一点处的连续性

为了定义函数在定义域 D 内的连续性，我们需要定义函数在一内点 x_0（即存在 $\delta > 0$，使得 $(x_0-\delta, x_0+\delta) \subset D$)）处的连续性.

定义 2.8 称函数 $f(x)$ 在其定义域的内点 x_0 处是**连续**的，如果
$$\lim_{x \to x_0} f(x) = f(x_0).$$

例1 图 2.10 给出了函数 $f(x)$ 的图像. 函数 $f(x)$ 在哪些点处是不连续的？为什么？

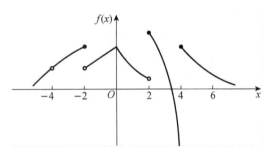

Figure 2.10
图 2.10

Solution $f(x)$ is discontinuous at $x=-4$ since $f(-4)$ is not defined. The graph also has a break when $x=-2$, $x=2$ and $x=4$. Here, $f(-2)$, $f(2)$ and $f(4)$ are defined, but $\lim_{x \to 2} f(x)$, $\lim_{x \to -2} f(x)$ and $\lim_{x \to 4} f(x)$ do not exist (because the left and right limits are different). So $f(x)$ is discontinuous at $x=2$, $x=-2$ and $x=4$.

From Definition 2.8, we see that a function $y=f(x)$ is continuous at a interior point $x=x_0$ of its domain if and only if it meets the following three conditions:

(1) $f(x_0)$ exists (x_0 lies in the domain of $f(x)$);
(2) $\lim_{x \to x_0} f(x)$ exists;
(3) $\lim_{x \to x_0} f(x) = f(x_0)$ (the limit equals the function value $f(x_0)$).

If any one of these three fails, then $f(x)$ is discontinuous at $x=x_0$, and x_0 is called a **discontinuity** of $f(x)$.

解 因为 $f(-4)$ 没有定义，所以 $f(x)$ 在 $x=-4$ 处是不连续的. 函数图像在 $x=-2$, $x=2$ 和 $x=4$ 时也间断. 这里，虽然 $f(-2)$, $f(2)$ 和 $f(4)$ 有定义，但是 $\lim_{x \to 2} f(x)$, $\lim_{x \to -2} f(x)$ 和 $\lim_{x \to 4} f(x)$ 不存在（因为在这些点处左、右极限不相等）. 因此，$f(x)$ 在点 $x=2$, $x=-2$ 和 $x=4$ 处是不连续的.

由定义 2.8 我们看出，函数 $y=f(x)$ 在其定义域的内点 $x=x_0$ 处是连续的当且仅当其满足下面三个条件：

(1) $f(x_0)$ 存在（x_0 在 $f(x)$ 的定义域中）；
(2) $\lim_{x \to x_0} f(x)$ 存在；
(3) $\lim_{x \to x_0} f(x) = f(x_0)$（极限等于其函数值 $f(x_0)$）.

如果三个条件中有一个不满足，那么 $f(x)$ 在点 $x=x_0$ 处是不连续的，并且 x_0 叫作 $f(x)$ 的一个**间断点**.

2. Several Common Types of Discontinuities

Example 2 $x = \frac{\pi}{2}$ is a discontinuity of the function $f(x) = \tan x$ since $f(x)$ is not defined at $x = \frac{\pi}{2}$. The discontinuity is called an **infinite discontinuity** of the function $\tan x$ since
$$\lim_{x \to \pi/2} \tan x = \infty.$$

Example 3 The function $y = \sin \frac{1}{x}$ is not defined at $x = 0$. As $x \to 0$, the values of the function oscillates back and forth between 1 and -1, so the limit of the function does not exist, see Figure 2.11. Thus, $x = 0$ is called an **oscillating discontinuity** of $y = \sin \frac{1}{x}$.

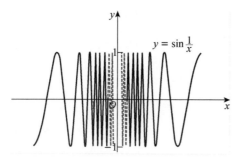

Figure 2.11
图 2.11

Example 4 The function $y = \frac{x^2 - 4}{x - 2}$ is not defined at $x = 2$. Thus, $x = 2$ is a discontinuity of the function (Figure 2.12). But
$$\lim_{x \to 2} \frac{x^2 - 4}{x - 2} = \lim_{x \to 2} (x + 2) = 4,$$
we could remove the discontinuity by setting $f(2) = 4$. Thus, this kind of discontinuity is called **removable discontinuity**.

Example 5 For the function
$$f(x) = \begin{cases} x - 1, & x < 0, \\ 0, & x = 0, \\ x + 1, & x > 0, \end{cases}$$

2. 间断点的几种常见类型

例 2 函数 $f(x) = \tan x$ 在点 $x = \frac{\pi}{2}$ 处没有定义,所以点 $x = \frac{\pi}{2}$ 是函数 $f(x) = \tan x$ 的间断点. 由于
$$\lim_{x \to \pi/2} \tan x = \infty,$$
我们称 $x = \frac{\pi}{2}$ 为函数 $\tan x$ 的**无穷间断点**.

例 3 函数 $y = \sin \frac{1}{x}$ 在点 $x = 0$ 处没有定义. 当 $x \to 0$ 时,函数值在 1 和 -1 之间来回振荡,所以函数的极限不存在,见图 2.11. 因此,$x = 0$ 叫作 $y = \sin \frac{1}{x}$ 的**振荡间断点**.

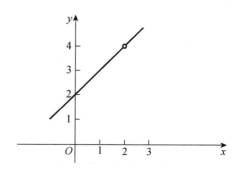

Figure 2.12
图 2.12

例 4 函数 $y = \frac{x^2 - 4}{x - 2}$ 在 $x = 2$ 处没有定义,因此 $x = 2$ 是函数的一个不连续点(图 2.12). 但是
$$\lim_{x \to 2} \frac{x^2 - 4}{x - 2} = \lim_{x \to 2} (x + 2) = 4,$$
我们可以通过令 $f(2) = 4$ 来消除该间断点. 因此,称这类间断点为**可去间断点**.

例 5 对于函数
$$f(x) = \begin{cases} x - 1, & x < 0, \\ 0, & x = 0, \\ x + 1, & x > 0, \end{cases}$$

we have
$$\lim_{x\to 0^-} f(x) = \lim_{x\to 0^-}(x-1) = -1,$$
$$\lim_{x\to 0^+} f(x) = \lim_{x\to 0^+}(x+1) = 1.$$

Since $\lim_{x\to 0^-} f(x) \neq \lim_{x\to 0^+} f(x)$, the limit $\lim_{x\to 0} f(x)$ does not exist. Thus $x=0$ is a discontinuity of the function $f(x)$. We called the discontinuity as a **jump discontinuity** because the graph of the function $f(x)$ "jumps" from -1 to 1 at $x=0$.

Definition 2.9 We say that a function $f(x)$ is **right-continuous** at the point x_0 if
$$\lim_{x\to x_0^+} f(x) = f(x_0).$$
And we say that $f(x)$ is **left-continuous** at the point x_0 if
$$\lim_{x\to x_0^-} f(x) = f(x_0).$$

From Definition 2.8 and Definition 2.9, we can see that a function is continuous at an interior point x_0 if and only if it is both right-continuous and left-continuous at x_0.

3. Continuity on an Interval

Definition 2.10 We say that a function $f(x)$ is **continuous** on an open interval (a,b) if it is continuous at each point of that interval. And we say that $f(x)$ is **continuous** on the closed interval $[a,b]$ if it is continuous on (a,b), right-continuous at a, and left-continuous at b.

The graph of continuous function is a continuous curve without breaking. From Definition 2.10, we see that a polynomial $f(x)$ is continuous on $(-\infty,+\infty)$ since for any number x_0,
$$\lim_{x\to x_0} f(x) = f(x_0).$$
And we also know that for rational function
$$F(x) = \frac{P(x)}{Q(x)},$$
when $Q(x_0) \neq 0$, $\lim_{x\to x_0} F(x) = F(x_0)$. Thus, the rational function is continuous at any point x_0 of its domain.

Example 6 Prove the absolute value function $f(x) = |x|$ is continuous on the interval $(-\infty,+\infty)$.

我们有
$$\lim_{x\to 0^-} f(x) = \lim_{x\to 0^-}(x-1) = -1,$$
$$\lim_{x\to 0^+} f(x) = \lim_{x\to 0^+}(x+1) = 1.$$

因为 $\lim_{x\to 0^-} f(x) \neq \lim_{x\to 0^+} f(x)$，所以极限 $\lim_{x\to 0} f(x)$ 不存在，从而 $x=0$ 是函数 $f(x)$ 的一个间断点. 我们称该间断点为**跳跃间断点**，因为函数 $f(x)$ 的图像在点 $x=0$ 处由 -1 跳跃到 1.

定义 2.9 如果
$$\lim_{x\to x_0^+} f(x) = f(x_0),$$
则称 $f(x)$ 在点 x_0 处**右连续**；如果
$$\lim_{x\to x_0^-} f(x) = f(x_0),$$
则称 $f(x)$ 在点 x_0 处**左连续**.

由定义 2.8 和定义 2.9，我们可以看出函数在内点 x_0 处连续当且仅当函数在点 x_0 处既右连续又左连续.

3. 区间上的连续性

定义 2.10 如果函数 $f(x)$ 在开区间 (a,b) 内的每一点都连续，则称 $f(x)$ 在开区间 (a,b) 内**连续**；如果函数 $f(x)$ 在 (a,b) 内连续，并且在点 a 处右连续，在点 b 处左连续，则称 $f(x)$ 在闭区间 $[a,b]$ 上**连续**.

连续函数的图像是一条连续不间断的曲线，由定义 2.10 可以看出，多项式 $f(x)$ 在 $(-\infty,+\infty)$ 上是连续的，因为对于每个数 x_0，有都有
$$\lim_{x\to x_0} f(x) = f(x_0).$$
我们也知道，对于有理函数
$$F(x) = \frac{P(x)}{Q(x)},$$
当 $Q(x_0) \neq 0$ 时，有 $\lim_{x\to x_0} F(x) = F(x_0)$. 因此，有理函数在其定义域内任何一点 x_0 处是连续的.

例 6 证明：绝对值函数 $f(x) = |x|$ 在区间 $(-\infty,+\infty)$ 内是连续的.

Proof The graph of the absolute value function $f(x)=|x|$ is shown in Figure 2.13.

For $x<0$, $f(x)=-x$, a polynomial; for $x>0$, $f(x)=x$, another polynomial. Thus $|x|$ is continuous at all numbers different from $x=0$. But $\lim\limits_{x\to 0}|x|=0=|0|$. Therefore, $|x|$ is also continuous at the point $x=0$. In summary, the absoute value function $f(x)$ is continuous on $(-\infty,+\infty)$.

证明 绝对值函数 $f(x)=|x|$ 的图像见图 2.13.

当 $x<0$ 时，$f(x)=-x$，是一个多项式；当 $x>0$ 时，$f(x)=x$，是另一个多项式. 因此，除了点 $x=0$，$|x|$ 都是连续的. 但是 $\lim\limits_{x\to 0}|x|=0=|0|$. 所以，$|x|$ 在点 $x=0$ 处也是连续的. 综上，绝对值函数 $f(x)=|x|$ 在 $(-\infty,+\infty)$ 内是连续的.

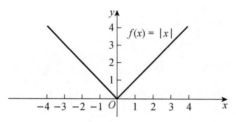

Figure 2.13
图 2.13

4. The Operations of Continuous Functions

Instead of using Definition 2.8, Definition 2.9 and Definition 2.10 to verify the continuity of a function, it is often convenient to use the next theorem, which shows how to build up complicated continuous functions from simple ones.

Theorem 2.3 If the functions $f(x)$ and $g(x)$ are continuous at x_0, and c is a constant, then the following functions are also continuous at x_0:

(1) $f+g$;   (2) $f-g$;   (3) cf;
(4) fg;   (5) $\dfrac{f}{g}$ ($g(x_0)\neq 0$).

Example 7 $\tan x=\dfrac{\sin x}{\cos x}, \cot x=\dfrac{\cos x}{\sin x}$, since $\sin x$ and $\cos x$ are continuous on the interval $(-\infty,+\infty)$, from theorem 2.3, $\tan x$ and $\cot x$ are continuous in their domain.

Theorem 2.4 If $f(u)$ is continuous at u_0 and $\lim\limits_{x\to x_0}g(x)=u_0$, then
$$\lim_{x\to x_0}f(g(x))=f(u_0).$$
In other words,
$$\lim_{x\to x_0}f(g(x))=f(\lim_{x\to x_0}g(x)).$$

4. 连续函数的运算

除了利用定义 2.8，定义 2.9 和定义 2.10 来证明函数的连续性，利用下面的定理，由简单的连续函数得到复杂的连续函数是很方便的.

定理 2.3 如果函数 $f(x)$ 和 $g(x)$ 在点 x_0 处连续，c 是一个常数，那么下列函数在点 x_0 处也是连续的：

(1) $f+g$;   (2) $f-g$;   (3) cf;
(4) fg;   (5) $\dfrac{f}{g}$ ($g(x_0)\neq 0$).

例 7 $\tan x=\dfrac{\sin x}{\cos x}, \cot x=\dfrac{\cos x}{\sin x}$. 由于 $\sin x$ 和 $\cos x$ 在区间 $(-\infty,+\infty)$ 上是连续的，由定理 2.3 知，$\tan x$ 和 $\cot x$ 在它们各自的定义域中都是连续的.

定理 2.4 如果函数 $f(u)$ 在点 u_0 处连续，并且 $\lim\limits_{x\to x_0}g(x)=u_0$，那么
$$\lim_{x\to x_0}f(g(x))=f(u_0).$$
换句话说，
$$\lim_{x\to x_0}f(g(x))=f(\lim_{x\to x_0}g(x)).$$

This theorem says that a limit symbol can be moved through a function symbol if the function $f(x)$ is continuous and the limit of $g(x)$ exists. In other words, the order of these two symbols can be reversed.

Example 8 Find $\lim\limits_{x\to 3}\sqrt{\dfrac{x-3}{x^2-9}}$.

Solution $\lim\limits_{x\to 3}\sqrt{\dfrac{x-3}{x^2-9}}=\sqrt{\lim\limits_{x\to 3}\dfrac{x-3}{x^2-9}}=\sqrt{\dfrac{1}{6}}=\dfrac{\sqrt{6}}{6}$.

Theorem 2.5 (Continuity of Composite Functions)

If $g(x)$ is continuous at x_0 and $g(x_0)=u_0$, $f(u)$ is continuous at $u=u_0$, then the composite $y=f(g(x))$ is continuous at x_0.

Proof Since $g(x)$ is continuous at x_0, we have
$$\lim_{x\to x_0} g(x) = g(x_0).$$
Since $f(u)$ is continuous at $u_0 = g(x_0)$, we can apply Theorem 2.4 to obtain
$$\lim_{x\to x_0} f(g(x)) = f(g(x_0)),$$
which is precisely the statement that the function $y=f(g(x))$ is continuous at x_0.

Example 9 Discuss the continuity of $y=\sin\dfrac{1}{x}$.

Solution Notice that the function $y=\sin\dfrac{1}{x}$ can be broken up as the composition of two functions:
$$y=\sin u \quad \text{and} \quad u=\dfrac{1}{x}.$$
$y=\sin u$ is continuous on the interval $(-\infty,+\infty)$; $u=\dfrac{1}{x}$ is continuous on the interval $(-\infty,0)\cup(0,+\infty)$. From Theorem 2.5, the function $y=\sin\dfrac{1}{x}$ is continuous on the interval $(-\infty,0)\cup(0,+\infty)$.

5. The Properties of Continuous Functions on a Closed Interval

Definition 2.11 Let I, the domain of $f(x)$, contain the point c.

2.5 The Continuity of Functions
2.5 函数的连续性

这个定理说明了如果函数 $f(x)$ 是连续的,并且 $g(x)$ 的极限存在,那么极限符号可以穿过函数符号.换句话说,这两个符号的顺序可以被交换.

例 8 求 $\lim\limits_{x\to 3}\sqrt{\dfrac{x-3}{x^2-9}}$.

解 $\lim\limits_{x\to 3}\sqrt{\dfrac{x-3}{x^2-9}}=\sqrt{\lim\limits_{x\to 3}\dfrac{x-3}{x^2-9}}=\sqrt{\dfrac{1}{6}}=\dfrac{\sqrt{6}}{6}$.

定理 2.5 (复合函数的连续性)

如果函数 $g(x)$ 在点 x_0 处连续并且 $g(x_0)=u_0$,函数 $f(u)$ 在点 $u=u_0$ 处连续,那么复合函数 $y=f(g(x))$ 在点 x_0 处连续.

证明 因为 $g(x)$ 在点 x_0 处是连续的,所以我们有
$$\lim_{x\to x_0} g(x) = g(x_0).$$
由于 $f(u)$ 在点 $u_0=g(x_0)$ 处是连续的,我们可以应用定理 2.4 得到
$$\lim_{x\to x_0} f(g(x)) = f(g(x_0)).$$
这精确地说明了函数 $y=f(g(x))$ 在点 x_0 处是连续的.

例 9 讨论 $y=\sin\dfrac{1}{x}$ 的连续性.

解 注意到函数 $y=\sin\dfrac{1}{x}$ 可以分解为两个函数
$$y=\sin u \quad \text{和} \quad u=\dfrac{1}{x}.$$
$y=\sin u$ 在区间 $(-\infty,+\infty)$ 内是连续的,而 $u=\dfrac{1}{x}$ 在区间 $(-\infty,0)\cup(0,+\infty)$ 内是连续的.由定理 2.5 可知,函数 $y=\sin\dfrac{1}{x}$ 在区间 $(-\infty,0)\cup(0,+\infty)$ 内是连续的.

5. 闭区间上连续函数的性质

定义 2.11 设在函数 f 的定义域 I 中存在一点 c.

(1) $f(c)$ is called the **maximum value** of $f(x)$ on I if $f(c) \geqslant f(x)$ for all x in I.

(2) $f(c)$ is called the **minimum value** of $f(x)$ on I if $f(c) \leqslant f(x)$ for all x in I.

The maximum value and the minimum value are collectively called **extreme value**.

The important properties of the continuous functions are expressed by the following theorems:

Theorem 2.6 (The Max-Min Existence Theorem)

If the function $f(x)$ is continuous on a closed interval $[a,b]$, then $f(x)$ attains both a maximum value and a minimum value there.

Inference If the function $f(x)$ is continuous on a closed interval $[a,b]$, then $f(x)$ is bounded on the closed interval.

An important property of continuous functions is expressed by the following theorem:

Theorem 2.7 (The Intermediate Value Theorem) Suppose that $f(x)$ is continuous on the closed interval $[a,b]$ and let C be any number between $f(a)$ and $f(b)$, where $f(a) \neq f(b)$. There exists a number $\xi \in (a,b)$, such that
$$f(\xi) = C.$$

The Intermediate Value Theorem states that a continuous function takes on every intermediate value between the function values $f(a)$ and $f(b)$. It is illustrated by Figure 2.14. Note that the value C can be taken on once or more than once.

(1) 如果对于 I 上的所有 x 有 $f(c) \geqslant f(x)$，则称 $f(c)$ 是在 I 上的**最大值**；

(2) 如果对于 I 上的所有 x 有 $f(c) \leqslant f(x)$，则称 $f(c)$ 是在 I 上的**最小值**.

最大值和最小值统称为**最值**.

下面的定理给出连续函数的重要性质：

定理 2.6(最大值-最小值存在定理)

如果函数 $f(x)$ 在闭区间 $[a,b]$ 上连续，那么 $f(x)$ 在该区间上存在最大值和最小值.

推论 如果函数 $f(x)$ 在闭区间 $[a,b]$ 上连续，那么 $f(x)$ 在该区间上有界.

下面的定理给出连续函数的一个重要性质：

定理 2.7(介值定理) 假设函数 $f(x)$ 在闭区间 $[a,b]$ 上连续，并且 C 为介于 $f(a)$ 和 $f(b)$ 之间的任何数，其中 $f(a) \neq f(b)$，那么一定存在数 $\xi \in (a,b)$，使得
$$f(\xi) = C.$$

介值定理表明，一个在区间 $[a,b]$ 上的连续函数 $f(x)$ 可以取到介于函数值 $f(a)$ 和 $f(b)$ 之间的任何值，见图 2.14，注意到 C 可以取到一次或多次.

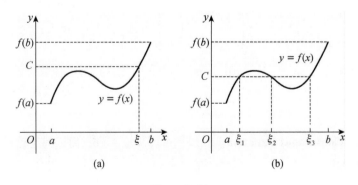

Figure 2.14
图 2.14

We can use the Intermediate Value Theorem to locate roots of equations.

Example 10 Show that there is a root of the equation
$$x^3 - 4x^2 + 1 = 0$$
in the interval $(0,1)$.

Proof Let
$$f(x) = x^3 - 4x^2 + 1.$$
It is easy to see that $f(x)$ is continuous on $[0,1]$.

We take $a=0, b=1, C=0$ in the Intermediate Value Theorem, we have
$$f(0) = 1 > 0, \quad f(1) = -2 < 0.$$
So there is a number ξ between 0 and 1 such that $f(\xi) = 0$, that is the equation $x^3 - 4x^2 + 1 = 0$ has at least one root ξ in the interval $(0,1)$.

我们可以利用介值定理来确定方程根的位置.

例 10 证明：方程
$$x^3 - 4x^2 + 1 = 0$$
在区间$(0,1)$内有一个根.

证明 令函数
$$f(x) = x^3 - 4x^2 + 1.$$
易发现$f(x)$在$[0,1]$上是连续的.

在介值定理中取$a=0, b=1, C=0$，我们有
$$f(0) = 1 > 0, \quad f(1) = -2 < 0.$$
因此，在 0 和 1 之间存在数ξ，使得$f(\xi) = 0$，即方程$x^3 - 4x^2 + 1 = 0$在区间$(0,1)$内至少有一个根ξ.

2.6 Infinitesimals and Infinitys
2.6 无穷小量和无穷大量

1. Infinitesimals

In this section, the symbol $\lim f(x)$ means a limit of the function $f(x)$ in some process of limit, the process may be $x \to x_0$, $x \to \infty$, etc.

Definition 2.12 If $\lim f(x) = 0$, then we say that $f(x)$ is an **infinitesimal** in the process of limit.

For example, since $\lim\limits_{x \to 0} x^2 = 0$, x^2 is an infinitesimal as $x \to 0$; Since $\lim\limits_{x \to \infty} \dfrac{1}{x^2} = 0$, $\dfrac{1}{x^2}$ is an infinitesimal as $x \to \infty$.

We can gain the following properties of infinitesimal by the definition of infinitesimal:

Property 1 If both of α, β are infinitesimals in the same process of limit, then $\alpha + \beta$, $\alpha - \beta$, $\alpha\beta$ still are infinitesimals in the process.

Property 2 The product of an infinitesimal and a bounded quantity is an infinitesimal.

Example 1 Find $\lim\limits_{x \to 0} x \sin \dfrac{1}{x}$.

1. 无穷小量

在本节中，符号$\lim f(x)$指的是某一极限过程中函数$f(x)$的极限，这个过程可能是$x \to x_0$, $x \to \infty$，等等.

定义 2.12 如果$\lim f(x) = 0$，则称$f(x)$是这个极限过程的**无穷小量**.

例如，因为$\lim\limits_{x \to 0} x^2 = 0$，所以函数$x^2$是$x \to 0$时的无穷小量；因为$\lim\limits_{x \to \infty} \dfrac{1}{x^2} = 0$，所以$\dfrac{1}{x^2}$为$x \to \infty$时的无穷小量.

由无穷小量的定义，可立刻得到下列性质：

性质 1 如果α, β为同一极限过程的无穷小量，则$\alpha + \beta$, $\alpha - \beta$, $\alpha\beta$仍是该极限过程的无穷小量.

性质 2 无穷小量与有界量的乘积是无穷小量.

例 1 求$\lim\limits_{x \to 0} x \sin \dfrac{1}{x}$.

Solution x is an infinitesimal as $x \to 0$. $\sin\dfrac{1}{x}$ is bounded. So
$$\lim_{x \to 0} x \sin\dfrac{1}{x} = 0.$$

2. Infinitys

Definition 2.13 If $\lim f(x) = \infty$, then we say that $f(x)$ is an **infinity** in the process of limit.

For example, $\lim\limits_{x \to 0}\dfrac{1}{x} = \infty$, so $\dfrac{1}{x}$ is an infinity as $x \to 0$.

Theorem 2.8 If $f(x)$ is an infinity in some process of limit, then $\dfrac{1}{f(x)}$ is an infinitesimal in the process. Conversely, if $f(x)$ is an infinitesimal and $f(x) \neq 0$ in some process of limit, then $\dfrac{1}{f(x)}$ is an infinity in the process.

Example 2 Find $\lim\limits_{x \to 1}\dfrac{2x-3}{x^2-5x+4}$.

Solution Since
$$\lim_{x \to 1}\dfrac{x^2-5x+4}{2x-3} = \dfrac{1^2-5 \cdot 1+4}{2 \cdot 1-3} = 0,$$
by the relation of infinitesimal and infinity, we have
$$\lim_{x \to 1}\dfrac{2x-3}{x^2-5x+4} = \infty.$$

3. Compare of Infinitesimals

We know that the sum, difference and product of two infinitesimals are still infinitesimals, but the quotient is not determined. For example,
$$\lim_{x \to 0}\dfrac{x^2}{2x} = 0, \quad \lim_{x \to 0}\dfrac{2x}{x^2} = \infty,$$
$$\lim_{x \to 0}\dfrac{\sin x}{x} = 1.$$

The infinitesimals are functions and their limits are zero, but their speed of convergence zero are different. We can estimate the speed through considering the limit of their quotient.

解 x 是 $x \to 0$ 时的无穷小量，$\sin\dfrac{1}{x}$ 是有界量，所以
$$\lim_{x \to 0} x \sin\dfrac{1}{x} = 0.$$

2. 无穷大量

定义 2.13 如果 $\lim f(x) = \infty$，则称 $f(x)$ 是这个极限过程的**无穷大量**.

例如，因为 $\lim\limits_{x \to 0}\dfrac{1}{x} = \infty$，所以 $\dfrac{1}{x}$ 是 $x \to 0$ 时的无穷大量.

定理 2.8 在某一极限变化过程中，若 $f(x)$ 为无穷大量，则 $\dfrac{1}{f(x)}$ 为无穷小量；反之，若 $f(x)$ 为无穷小量，且 $f(x) \neq 0$，则 $\dfrac{1}{f(x)}$ 为无穷大量.

例 2 求 $\lim\limits_{x \to 1}\dfrac{2x-3}{x^2-5x+4}$.

解 由于
$$\lim_{x \to 1}\dfrac{x^2-5x+4}{2x-3} = \dfrac{1^2-5 \cdot 1+4}{2 \cdot 1-3} = 0,$$
根据无穷大量与无穷小量的关系，得
$$\lim_{x \to 1}\dfrac{2x-3}{x^2-5x+4} = \infty.$$

3. 无穷小量的比较

两个无穷小量的和、差、积仍为无穷小量，但其商未必是无穷小量. 例如
$$\lim_{x \to 0}\dfrac{x^2}{2x} = 0, \quad \lim_{x \to 0}\dfrac{2x}{x^2} = \infty,$$
$$\lim_{x \to 0}\dfrac{\sin x}{x} = 1.$$

无穷小量是以 0 为极限的函数，然而不同的无穷小量收敛于 0 的速度有快慢之分. 我们通过考查两个无穷小量的比值的极限，来判断两个无穷小量收敛速度的快慢.

Suppose that α and β are both infinitesimals in a same process of limit, and α≠0.

Definition 2.14 If $\lim\dfrac{\beta}{\alpha}=0$, then we say that β is an **infinitesimal of higher order** relative to α, write $\beta=o(\alpha)$.

If $\lim\dfrac{\beta}{\alpha}=\infty$, then we say that β is an **infinitesimal of lower order** relative to α.

If $\lim\dfrac{\beta}{\alpha}=c\neq 0$, then we say that β is an **infinitesimal of same order** relative to α.

If $\lim\dfrac{\beta}{\alpha^k}=c\neq 0, k>0$, then we say that β is an **infinitesimal of k order** relative to α.

If $\lim\dfrac{\beta}{\alpha}=1$, then we say that β is an **equivalent infinitesimal** relative to α, write $\alpha\sim\beta$.

For example, $\lim\limits_{x\to 0}\dfrac{2017x^2}{x}=0$, so $2017x^2=o(x)$ as $x\to 0$; $\lim\limits_{x\to 0}\dfrac{x^2}{x^3}=\infty$, so x^2 is an infinitesimal of lower order relative to x^3 as $x\to 0$; $\lim\limits_{x\to 0}\dfrac{\sin x}{x}=1$, so $\sin x$ is an equivalent infinitesimal relative to x as $x\to 0$, that is

$$\sin x \sim x \quad (x\to 0).$$

We can also know that

$$\tan x \sim x, \quad 1-\cos x \sim \dfrac{1}{2}x^2,$$

$$\arcsin x \sim x$$

as $x\to 0$.

Theorem 2.9 (The Equivalent Infinitesimal Theorem)
Suppose that $f(x), g(x), h(x)$ are infinitesimals in a same process of limit, and $f(x)\sim g(x)$. We have

(1) if $\lim f(x)h(x)=A$, then
$$\lim g(x)h(x)=A;$$

(2) if $\lim\dfrac{h(x)}{f(x)}=B$, then
$$\lim\dfrac{h(x)}{g(x)}=B.$$

下面假设 α 和 β 都是某个相同极限过程的无穷小量，且 α≠0.

定义 2.14 如果 $\lim\dfrac{\beta}{\alpha}=0$，则称 β 是比 α **高阶的无穷小量**，记作 $\beta=o(\alpha)$；

如果 $\lim\dfrac{\beta}{\alpha}=\infty$，则称 β 是比 α **低阶的无穷小量**；

如果 $\lim\dfrac{\beta}{\alpha}=c\neq 0$，则称 β 与 α 是**同阶无穷小量**；

如果 $\lim\dfrac{\beta}{\alpha^k}=c\neq 0, k>0$，则称 β 是关于 α 的 k **阶无穷小量**；

如果 $\lim\dfrac{\beta}{\alpha}=1$，则称 β 与 α 是**等价无穷小量**，记作 $\alpha\sim\beta$.

例如，因为 $\lim\limits_{x\to 0}\dfrac{2017x^2}{x}=0$，所以当 $x\to 0$ 时，$2017x^2=o(x)$；因为 $\lim\limits_{x\to 0}\dfrac{x^2}{x^3}=\infty$，所以当 $x\to 0$ 时，x^2 是比 x^3 低阶的无穷小量；因为 $\lim\limits_{x\to 0}\dfrac{\sin x}{x}=1$，所以当 $x\to 0$ 时，$\sin x$ 与 x 是等价无穷小量，即

$$\sin x \sim x \quad (x\to 0).$$

我们也不难知道，当 $x\to 0$ 时，

$$\tan x \sim x, \quad 1-\cos x \sim \dfrac{1}{2}x^2,$$

$$\arcsin x \sim x.$$

定理 2.9（无穷小量等价定理）
设函数 $f(x), g(x), h(x)$ 都是某个相同极限过程的无穷小量，且 $f(x)\sim g(x)$，则有

(1) 如果 $\lim f(x)h(x)=A$，那么
$$\lim g(x)h(x)=A;$$

(2) 如果 $\lim\dfrac{h(x)}{f(x)}=B$，那么
$$\lim\dfrac{h(x)}{g(x)}=B.$$

Example 3 Find $\lim\limits_{x\to 0}\dfrac{1-\cos x}{\sin^2 4x}$.

Solution $\lim\limits_{x\to 0}\dfrac{1-\cos x}{\sin^2 4x}=\lim\limits_{x\to 0}\dfrac{\frac{1}{2}x^2}{(4x)^2}=\dfrac{1}{32}.$

Example 4 Find $\lim\limits_{x\to 0}\dfrac{\sin x}{\arcsin 2x}$.

Solution $\lim\limits_{x\to 0}\dfrac{\sin x}{\arcsin 2x}=\lim\limits_{x\to 0}\dfrac{x}{2x}=\dfrac{1}{2}.$

例 3 求 $\lim\limits_{x\to 0}\dfrac{1-\cos x}{\sin^2 4x}$.

解 $\lim\limits_{x\to 0}\dfrac{1-\cos x}{\sin^2 4x}=\lim\limits_{x\to 0}\dfrac{\frac{1}{2}x^2}{(4x)^2}=\dfrac{1}{32}.$

例 4 求 $\lim\limits_{x\to 0}\dfrac{\sin x}{\arcsin 2x}$.

解 $\lim\limits_{x\to 0}\dfrac{\sin x}{\arcsin 2x}=\lim\limits_{x\to 0}\dfrac{x}{2x}=\dfrac{1}{2}.$

Exercises 2
习题 2

1. Find the limits of the following sequences if they exist:

(1) $a_n=\dfrac{1}{2^n}, n=1,2,\cdots$;

(2) $a_n=(-1)^n\dfrac{1}{n}, n=1,2,\cdots$;

(3) $a_n=2+\dfrac{1}{n^2}, n=1,2,\cdots$;

(4) $a_n=\dfrac{n-1}{n+1}, n=1,2,\cdots$;

(5) $a_n=n(-1)^n, n=1,2,\cdots$.

2. Evaluate the following limits if they exist:

(1) $\lim\limits_{x\to -2}(3x^2-5x+1)$;

(2) $\lim\limits_{x\to\sqrt{3}}\dfrac{x^2-3}{x^4+x^2+1}$;

(3) $\lim\limits_{x\to 0}\left(1-\dfrac{2}{x-3}\right)$;

(4) $\lim\limits_{x\to 2}\dfrac{x^2-3}{x-3}$;

(5) $\lim\limits_{x\to 1}\dfrac{x^2-1}{2x^2-x-1}$;

(6) $\lim\limits_{x\to 0}\dfrac{4x^3-2x^2+x}{3x^2+2x}$;

(7) $\lim\limits_{x\to\infty}\dfrac{3x+2}{6x-1}$;

(8) $\lim\limits_{x\to\infty}\dfrac{500x}{1+x^3}$;

(9) $\lim\limits_{x\to\infty}\dfrac{(2x+1)^{30}(3x-2)^{20}}{(5x+3)^{50}}$;

1. 如果下列数列的极限存在, 请求出:

(1) $a_n=\dfrac{1}{2^n}, n=1,2,\cdots$;

(2) $a_n=(-1)^n\dfrac{1}{n}, n=1,2,\cdots$;

(3) $a_n=2+\dfrac{1}{n^2}, n=1,2,\cdots$;

(4) $a_n=\dfrac{n-1}{n+1}, n=1,2,\cdots$;

(5) $a_n=n(-1)^n, n=1,2,\cdots$.

2. 如果下列极限存在, 请求出:

(1) $\lim\limits_{x\to -2}(3x^2-5x+1)$;

(2) $\lim\limits_{x\to\sqrt{3}}\dfrac{x^2-3}{x^4+x^2+1}$;

(3) $\lim\limits_{x\to 0}\left(1-\dfrac{2}{x-3}\right)$;

(4) $\lim\limits_{x\to 2}\dfrac{x^2-3}{x-3}$;

(5) $\lim\limits_{x\to 1}\dfrac{x^2-1}{2x^2-x-1}$;

(6) $\lim\limits_{x\to 0}\dfrac{4x^3-2x^2+x}{3x^2+2x}$;

(7) $\lim\limits_{x\to\infty}\dfrac{3x+2}{6x-1}$;

(8) $\lim\limits_{x\to\infty}\dfrac{500x}{1+x^3}$;

(9) $\lim\limits_{x\to\infty}\dfrac{(2x+1)^{30}(3x-2)^{20}}{(5x+3)^{50}}$;

(10) $\lim\limits_{x\to 3}\dfrac{x^2-5x+6}{x^2-8x+15}$;

(11) $\lim\limits_{x\to\infty}\dfrac{x^4-8x+1}{x^2+3}$;

(12) $\lim\limits_{x\to 3}\dfrac{5x^2-7x-24}{x^2+2}$;

(13) $\lim\limits_{x\to 1/4}\dfrac{x^3-2x^2+5x-1}{3x^3-2}$;

(14) $\lim\limits_{x\to 7}\dfrac{\sqrt{x+2}-3}{x-7}$;

(15) $\lim\limits_{t\to 9}\dfrac{9-t}{3-\sqrt{t}}$.

3. Find the following limits of the functions:

(1) $\lim\limits_{x\to 0}\dfrac{\sin 3x}{\sin 5x}$;

(2) $\lim\limits_{x\to 0}\dfrac{\tan 2x-\sin x}{x}$;

(3) $\lim\limits_{x\to 0}\dfrac{\cos x-\cos 3x}{x^2}$;

(4) $\lim\limits_{x\to 0}\dfrac{\tan(2x+x^3)}{\sin(x-x^2)}$;

(5) $\lim\limits_{x\to\infty}x\sin\dfrac{2}{x}$;

(6) $\lim\limits_{x\to 0}\dfrac{x-\sin x}{x+\sin x}$.

4. Find the following limits of the functions:

(1) $\lim\limits_{x\to 0}\left(\dfrac{3-x}{3}\right)^{\frac{2}{x}}$;

(2) $\lim\limits_{x\to\infty}\left(\dfrac{x-1}{x+1}\right)^{x}$;

(3) $\lim\limits_{x\to\infty}\left(1+\dfrac{4}{x}\right)^{2x}$;

(4) $\lim\limits_{x\to\infty}\left(1-\dfrac{2}{x}\right)^{\frac{x}{2}-1}$.

5. Where are each of the following functions discontinuous? Why?

(1) $y=\dfrac{1}{(x+3)^2}$;

(2) $y=x\cos\dfrac{1}{x}$;

(3) $y=\dfrac{x^3-1}{x^2-1}$;

(4) $y=(1+x)^{\frac{1}{x}}$.

6. Use continuity to evaluate the following limits:

(1) $\lim\limits_{x\to 0}\sqrt{1+2x-x^2}$;

(2) $\lim\limits_{x\to 0}\dfrac{\cot(1+x)}{\cos(1+x^2)}$;

(3) $\lim\limits_{x\to 0}\left[\dfrac{\lg(100+x)}{2^x+\tan x}\right]^{\frac{1}{2}}$;

(4) $\lim\limits_{x\to 1}\arctan\sqrt{\dfrac{x^2+1}{x+1}}$.

7. Show that the function $f(x)=1-\sqrt{1-x^2}$ is continuous on the interval $[-1,1]$.

8. For what value of the constant a is the function
$$f(x)=\begin{cases}e^x, & x<0,\\ a+x, & x\geq 0\end{cases}$$
continuous on $(-\infty,+\infty)$?

9. Show that there is a root of the equation
$$x^5-3x=1$$
between 1 and 2.

10. Compute the following limits:

(1) $\lim\limits_{x\to\infty}\dfrac{2x+1}{x}$;

(2) $\lim\limits_{x\to 0}\dfrac{1-x^2}{1-x}$;

(3) $\lim\limits_{h\to 0}\dfrac{(x+h)^2-x^2}{h}$;

(4) $\lim\limits_{x\to 0}\dfrac{\tan 3x}{2x}$;

(5) $\lim\limits_{x\to 0}\dfrac{\sin x^n}{(\sin x)^m}$ (n,m are both integer numbers);

(6) $\lim\limits_{x\to 0}\dfrac{\tan x-\sin x}{\sin^3 x}$.

Chapter 3 The Derivative and the Differential
第 3 章 导数和微分

Differential integral is an important part in Calculus, the basic concept of differential integral are derivative and differential.

In this chapter, we mainly discuss the the basic concept of derivative and differential, the calculation methods. For the application of derivative, we will discuss it in Chapter 4.

微分学是微积分的重要组成部分,它的基本概念是导数与微分.

在本章中,我们主要讨论导数和微分的概念以及它们的计算方法. 至于导数的应用,我们将在第 4 章中讨论.

3.1 The Concept of the Derivative
3.1 导数的概念

1. Introducing Examples

1) Average Velocity and Instantaneous Velocity Problems

We consider the example of an object falling in a vacuum. This experiment shows that if it starts from rest, object falls $16t^2$ ft in t seconds, and 64 ft during the first 2 seconds; clearly, it falls faster and faster as time goes on, the distance travels as a function of time. During the second second (from $t=1$ to $t=2$), the object falls 48 ft. So, its average velocity is

$$v_{\text{avg}} = \frac{64-16}{2-1} \text{ ft/s} = 48 \text{ ft/s}.$$

During the time interval from $t=1$ to $t=1.5$, it falls 16×1.5^2 ft $- 16 \times 1^2$ ft $= 20$ feet, its average velocity is

$$v_{\text{avg}} = \frac{16 \times 1.5^2 - 16 \times 1^2}{1.5-1} \text{ ft/s}$$
$$= \frac{20}{0.5} \text{ ft/s} = 40 \text{ ft/s}.$$

Similarly, on the time intervals $t=1$ to $t=1.01$ and $t=1$ to $t=1.001$, the respective average velocities are

1. 引例

1) 平均速度和瞬时速度问题

我们考虑一个物体在真空中下落的例子. 试验表明,如果物体是从静止开始下落,在 t 秒内下落了 $16t^2$ ft,在前两秒内下落了 64 ft;随着时间的推移,物体下落得越来越快,下落的距离就是时间的函数. 所以,在第 2 秒(从 $t=1$ 到 $t=2$)内,物体下落了 48 ft,其下落的平均速度是

$$v_{\text{avg}} = \frac{64-16}{2-1} \text{ ft/s} = 48 \text{ ft/s}.$$

在 $t=1$ s 到 $t=1.5$ s 时间段内,它下降了 16×1.5^2 ft $- 16 \times 1^2$ ft $= 20$ ft,其平均速度是

$$v_{\text{avg}} = \frac{16 \times 1.5^2 - 16 \times 1^2}{1.5-1} \text{ ft/s}$$
$$= \frac{20}{0.5} \text{ ft/s} = 40 \text{ ft/s}.$$

类似地,在 $t=1$ 到 $t=1.01$ 以及 $t=1$ 到 $t=1.001$ 时间段内,相应的平均速度是

$$v_{\text{avg}}=\frac{16\times1.01^2-16\times1^2}{1.01-1}\ \text{ft/s}=\frac{0.3216}{0.01}\ \text{ft/s}=32.16\ \text{ft/s},$$

$$v_{\text{avg}}=\frac{16\times1.001^2-16\times1^2}{1.001-1}\ \text{ft/s}=\frac{0.032\,016}{0.001}\ \text{ft/s}=32.016\ \text{ft/s}.$$

$$v_{\text{avg}}=\frac{16\times1.01^2-16\times1^2}{1.01-1}\ \text{ft/s}=\frac{0.3216}{0.01}\ \text{ft/s}=32.16\ \text{ft/s},$$

$$v_{\text{avg}}=\frac{16\times1.001^2-16\times1^2}{1.001-1}\ \text{ft/s}=\frac{0.032\,016}{0.001}\ \text{ft/s}=32.016\ \text{ft/s}.$$

How do we calculate the instantaneous velocity at the instant $t=1$. The shorter the time interval is, the better we should approximate the instantaneous velocity at the instant $t=1$. Looking at the numbers 48, 40, 32.16, 32.016, we guess that 32 ft/s to be the instantaneous velocity at the instant $t=1$.

如何计算 $t=1$ 时的瞬时速度呢？时间段越短，计算出的速度就越接近 $t=1$ 时的瞬时速度. 从数字 48，40，32.16，32.016，可以猜测 $t=1$ 时的瞬时速度是 32 ft/s.

We can describe this problem more accurately that an object moves along a coordinate line so that its position at time t is given by $s=f(t)$. At time a the object is at $f(a)$; at the nearby time $a+h$, it is at $f(a+h)$. Thus, the average velocity on the this interval from $[a,a+h]$ is

$$v_{\text{avg}}=\frac{f(a+h)-f(a)}{h}.$$

The instantaneous velocity at time a is

$$v=\lim_{h\to0}v_{\text{avg}}=\lim_{h\to0}\frac{f(a+h)-f(a)}{h}$$

(provided that the limit exists).

我们把这个问题更准确地描述为：物体沿着坐标轴移动，它在时刻 t 的位置是 $s=f(t)$. 在时刻 a，物体位于 $f(a)$；在邻近的时刻 $a+h$，物体的位置是 $f(a+h)$. 因此，在时间段$[a,a+h]$内的平均速度是

$$v_{\text{avg}}=\frac{f(a+h)-f(a)}{h}.$$

在时刻 a 的瞬时速度是

$$v=\lim_{h\to0}v_{\text{avg}}=\lim_{h\to0}\frac{f(a+h)-f(a)}{h}$$

（假设此极限存在）.

2) Tangency Problems

2）切线问题

For circles, tangency is straightforward. A line L is tangent to a circle at a point P if L passes through P, and perpendicular to the radius at P (Figure 3.1). Such a line just touches the circle. But what does it mean to say that a line L is tangent to some other curve C at a point P?

对于圆来说，相切的概念是很直接的. 直线 L 是圆周上一点 P 的切线，如果 L 过点 P 且垂直于点 P 的半径（图 3.1）. 这样一条直线刚好触到圆周. 但是，说直线 L 是另一条曲线 C 在点 P 处的切线的含义是什么呢？

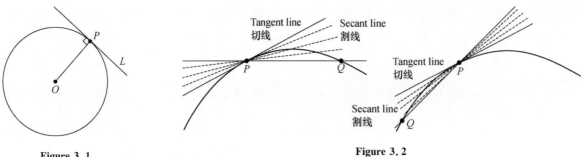

Figure 3.1

图 3.1

Figure 3.2

图 3.2

To define tangency to an arbitrary curve at a point, we apply a dynamic method to observe the state of the secant line PQ passing through P and a nearby point Q, where Q moves forward P along the curve (Figure 3.2). The procedure is as follows:

(1) Calculate the slope of the secant line PQ.

(2) Consider the limit of the slop of the secant line PQ with Q tending toward P along the curve.

(3) Provide the limit exists, we call it the slope of tangent line of the curve at P and define the tangent line of the curve at P to be the line passing through P with this slope.

Generally, the slope of a tangent line of an arbitrary curve $y=f(x)$ at the point $P(x_0,f(x_0))$ is

$$k=\lim_{h\to 0}\frac{f(x_0+h)-f(x_0)}{h}$$

(provide the limit exists). So the tangent line of the curve $y=f(x)$ at P is the line through P with this slope k (Figure 3.3).

为了定义与一般曲线在一点相切的概念，我们需要一种动态处理的方法. 这种方法考虑了过点 P 和附近点 Q 的割线 PQ 当点 Q 沿曲线向点 P 移动时的形态（图 3.2）. 该方法的大致步骤如下：

(1) 计算割线 PQ 的斜率.

(2) 考虑当点 Q 沿曲线趋于点 P 时割线 PQ 的斜率的极限.

(3) 如果极限存在，就把它取作曲线在点 P 处切线的斜率，并把过点 P 且具有这个斜率的直线定义为曲线在点 P 处的切线.

一般地，任意曲线 $y=f(x)$ 在点 $P(x_0,f(x_0))$ 处切线的斜率是

$$k=\lim_{h\to 0}\frac{f(x_0+h)-f(x_0)}{h}$$

（倘若这个极限存在）. 于是，曲线 $y=f(x)$ 上点 P 处的切线是斜率为 k 的且过点 P 的直线（图 3.3）.

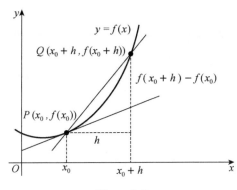

Figure 3.3

图 3.3

2. The Derivative of Function at a Point

The two examples above are both boiled down to find the limit of the following form:

$$\begin{aligned}f'(a)&=\lim_{h\to 0}\frac{f(a+h)-f(a)}{h}\\&=\lim_{\Delta x\to 0}\frac{f(a+\Delta x)-f(a)}{\Delta x}\\&=\lim_{\Delta x\to 0}\frac{\Delta y}{\Delta x}.\end{aligned}$$

2. 函数在一点处的导数

上面两个例子最终都归结为求如下形式的极限：

$$\begin{aligned}f'(a)&=\lim_{h\to 0}\frac{f(a+h)-f(a)}{h}\\&=\lim_{\Delta x\to 0}\frac{f(a+\Delta x)-f(a)}{\Delta x}\\&=\lim_{\Delta x\to 0}\frac{\Delta y}{\Delta x}.\end{aligned}$$

In fact, limits of this form arise whenever we calculate a rate of change in any of the sciences or engineering, such as a rate of reaction in chemistry or a marginal cost in economics. Since this type of limit occurs so widely, it is given a special name and notation.

Definition 3.1 Provided the function $y=f(x)$ is definited in the internal which is including $x=a$. If the limit

$$f'(a)=\lim_{\Delta x\to 0}\frac{\Delta y}{\Delta x}=\lim_{\Delta x\to 0}\frac{f(a+\Delta x)-f(a)}{\Delta x}$$

or

$$f'(a)=\lim_{h\to 0}\frac{f(a+h)-f(a)}{h}$$

exists, we called the limit as the **derivative** of the function $f(x)$ at $x=a$, denoted by $f'(a)$. Now we also say that $f(x)$ is **differentiable** at $x=a$. We can write $f'(a)$ as

$$y'\big|_{x=a},\quad \frac{\mathrm{d}y}{\mathrm{d}x}\big|_{x=a},\quad \text{or}\quad \frac{\mathrm{d}f(x)}{\mathrm{d}x}\big|_{x=a}.$$

If we write $x=a+\Delta x$, then we have $\Delta x=x-a$ and Δx approaches 0 if and only if x approaches a. Therefore, an equivalent way of stating the definition of the derivative is

$$f'(a)=\lim_{x\to a}\frac{f(x)-f(a)}{x-a}.$$

Example 1 Find the derivative of the function

$$f(x)=x^2-8x+9$$

at $x=a$.

Solution From Definition 3.1, we have

$$f'(a)=\lim_{h\to 0}\frac{f(a+h)-f(a)}{h}$$
$$=\lim_{h\to 0}\frac{[(a+h)^2-8(a+h)+9]-[a^2-8a+9]}{h}$$
$$=\lim_{h\to 0}\frac{a^2+2ah+h^2-8a-8h+9-a^2+8a-9}{h}$$
$$=\lim_{h\to 0}\frac{2ah+h^2-8h}{h}=\lim_{h\to 0}(2a+h-8)$$
$$=2a-8.$$

According to the introducing example, the geometry significant of derivation is as follow: the slope of the tangent line of the cure $y=f(x)$ at $(a,f(a))$ is equal to $f'(a)$. If we use the point-slope form of the equation of a line, then the equation of the tangent line of the curve $y=f(x)$ at the point $(a,f(a))$ is

事实上,在任何科学和工程中,只要计算变化率时,都归结为求这种类型的极限,例如化学中的反应速率、经济学中的边际收益等.由于这种类型的极限应用广泛,所以我们给它一个特殊的名字和记号.

定义 3.1 设函数 $y=f(x)$ 在包含点 $x=a$ 的区间内有定义.如果极限

$$f'(a)=\lim_{\Delta x\to 0}\frac{\Delta y}{\Delta x}=\lim_{\Delta x\to 0}\frac{f(a+\Delta x)-f(a)}{\Delta x}$$

或

$$f'(a)=\lim_{h\to 0}\frac{f(a+h)-f(a)}{h}$$

存在,则称此极限为函数 $f(x)$ 在点 $x=a$ 处的**导数**,记作 $f'(a)$.这时也称 $f(x)$ 在点 $x=a$ 处**可导**.我们也可把 $f'(a)$ 记为

$$y'\big|_{x=a},\quad \frac{\mathrm{d}y}{\mathrm{d}x}\big|_{x=a}\quad 或\quad \frac{\mathrm{d}f(x)}{\mathrm{d}x}\big|_{x=a}.$$

如果我们记 $x=a+\Delta x$,那么我们有 $\Delta x=x-a$,Δx 趋向于 0 当且仅当 x 趋向于 a.因此,导数定义的一个等价形式就是

$$f'(a)=\lim_{x\to a}\frac{f(x)-f(a)}{x-a}.$$

例 1 计算函数

$$f(x)=x^2-8x+9$$

在点 $x=a$ 处的导数.

解 利用定义 3.1,我们有

$$f'(a)=\lim_{h\to 0}\frac{f(a+h)-f(a)}{h}$$
$$=\lim_{h\to 0}\frac{[(a+h)^2-8(a+h)+9]-[a^2-8a+9]}{h}$$
$$=\lim_{h\to 0}\frac{a^2+2ah-8a-8h+9-a^2+8a\quad 9}{h}$$
$$=\lim_{h\to 0}\frac{2ah+h^2-8h}{h}=\lim_{h\to 0}(2a+h-8)$$
$$=2a-8.$$

根据引例知导数的几何意义是:$f'(a)$ 就是曲线 $y=f(x)$ 在点 $(a,f(a))$ 处的切线的斜率.如果我们采用点斜式直线方程,则曲线 $y=f(x)$ 在点 $(a,f(a))$ 处的切线方程为

$$y - f(a) = f'(a)(x-a).$$

Example 2 Find the equation of the tangent line of $y = x^2 - 2x + 3$ at the point $x = 2$.

Solution By the definition of derivative, we can compute the derivative of $y = x^2 - 2x + 3$ is
$$y' = 2x - 2.$$
If $x = 2$, then $y = 3$. the slope of the tangent line at $(2,3)$ is
$$y'|_{x=2} = 2 \times 2 - 2 = 2.$$
Thus the equation of the tangent line is
$$y - 3 = 2(x-2) \quad \text{or} \quad y = 2x - 3.$$

3. One-sided Derivatives

If the limit
$$\lim_{h \to 0^+} \frac{f(a+h) - f(a)}{h}$$
exists, it is called the **right-hand derivative** of the function $f(x)$ at a and denoted by $f'_+(a)$.

If the limit
$$\lim_{h \to 0^-} \frac{f(a+h) - f(a)}{h}$$
exists, it is called the **left-hand derivative** of the function $f(x)$ at a and denoted by $f'_-(a)$.

According to the sufficient and necessary condition of limit, a function has a derivative at a point if and only if it has the right-hand and the left-hand derivatives there and these one-sided derivatives are equal.

We say that the function $y = f(x)$ is **differentiable** on an open interval (finite or infinite), if it has a derivative at each point of the interval. The function $y = f(x)$ is differentiable on a closed interval $[a,b]$, if it is differentiable on the open interval (a,b) and if the limits $f'_+(a)$ and $f'_-(b)$ exist.

Example 3 Show that the function $f(x) = |x|$ is not differentiable at $x = 0$.

Proof The right-hand derivative of $f(x) = |x|$ at $x = 0$ is
$$f'_+(0) = \lim_{h \to 0^+} \frac{|0+h| - |0|}{h}$$
$$= \lim_{h \to 0^+} \frac{|h|}{h} = \lim_{h \to 0^+} \frac{h}{h} = 1.$$

例 2 求曲线 $y = x^2 - 2x + 3$ 在点 $x = 2$ 处的切线方程.

解 利用导数的定义,我们可以计算出函数 $y = x^2 - 2x + 3$ 的导数是
$$y' = 2x - 2.$$
当 $x = 2$ 时,$y = 3$. 点$(2,3)$处的切线斜率是
$$y'|_{x=2} = 2 \times 2 - 2 = 2.$$
于是所求的切线方程为
$$y - 3 = 2(x-2) \quad \text{或} \quad y = 2x - 3.$$

3. 单侧导数

如果极限
$$\lim_{h \to 0^+} \frac{f(a+h) - f(a)}{h}$$
存在,则称此极限为函数 $f(x)$ 在点 a 处的**右导数**,记为 $f'_+(a)$.

如果极限
$$\lim_{h \to 0^-} \frac{f(a+h) - f(a)}{h}$$
存在,则称此极限为函数 $f(x)$ 在点 a 处的**左导数**,记为 $f'_-(a)$.

由极限存在的充分必要条件知,函数在一点处可导当且仅当函数在该点处的左导数和右导数存在而且两者相等.

如果函数 $f(x)$ 在一开区间(有限或无限)中的每点处都可导,则称 $f(x)$ 在此区间内**可导**. 如果函数 $f(x)$ 在开区间(a,b)内可导,而且 $f'_+(a)$ 和 $f'_-(b)$ 存在,则称函数 $f(x)$ 在闭区间$[a,b]$上可导.

例 3 证明:函数 $f(x) = |x|$ 在点 $x = 0$ 处不可导.

证明 $f(x) = |x|$ 在点 $x = 0$ 处的右导数是
$$f'_+(0) = \lim_{h \to 0^+} \frac{|0+h| - |0|}{h}$$
$$= \lim_{h \to 0^+} \frac{|h|}{h} = \lim_{h \to 0^+} \frac{h}{h} = 1,$$

The left-hand derivative of $f(x)=|x|$ at $x=0$ is

$$f'_-(0)=\lim_{h\to 0^-}\frac{|0+h|-|0|}{h}$$
$$=\lim_{h\to 0^-}\frac{|h|}{h}=\lim_{h\to 0^-}\frac{-h}{h}=-1.$$

Since the right-hand derivative and the left-hand derivative are different, $f(x)=|x|$ is not differentiable at $x=0$.

4. The Derivative of a Function

If the function $f(x)$ is differentiable on the interval I, $f'(x_0)$ exits and is unique for any number x_0 in the interval I. Then we can get a function which is defined on the interval I and called the **derived function**. It is also called the **derivative** of $f(x)$ which is denoted by $f'(x)$. That is

$$f'(x)=\lim_{h\to 0}\frac{f(x+h)-f(x)}{h}.$$

The domain of $f'(x)$ is the set $\{x\,|\,f'(x)\text{ exists}\}$ and may be smaller than the domain of $f(x)$.

Example 4 Find the derivative of $f(x)=C$ (C is a constant).

Solution $f'(x)=\lim_{h\to 0}\dfrac{f(x+h)-f(x)}{h}$
$$=\lim_{h\to 0}\frac{C-C}{h}=0.$$

That is

$$(C)'=0.$$

Example 5 Find the derivative of $f(x)=x^n$ (n is positive integer) at $x=a$.

Solution $f'(a)=\lim_{x\to a}\dfrac{f(x)-f(a)}{x-a}$
$$=\lim_{x\to a}\frac{x^n-a^n}{x-a}$$
$$=\lim_{x\to a}(x^{n-1}+ax^{n-2}+\cdots+a^{n-1})$$
$$=na^{n-1}.$$

Replace a by x, we get

$$f'(x)=nx^{n-1}.$$

That is

$$(x^n)'=nx^{n-1}.$$

Generally, we have $(x^\mu)' = \mu x^{\mu-1}$, where μ is a constant. For example,

$$(C)' = 0, \quad \left(\frac{1}{x}\right)' = -\frac{1}{x^2},$$

$$(\sqrt{x})' = \frac{1}{2\sqrt{x}}, \quad (x^\mu)' = \mu x^{\mu-1}.$$

Example 6 Find the derivative of $f(x) = \sin x$.

Solution
$$\begin{aligned}
f'(x) &= \lim_{h \to 0} \frac{f(x+h) - f(x)}{h} \\
&= \lim_{h \to 0} \frac{\sin(x+h) - \sin x}{h} \\
&= \lim_{h \to 0} \frac{1}{h} \cdot 2\cos\left(x + \frac{h}{2}\right) \sin \frac{h}{2} \\
&= \lim_{h \to 0} \cos\left(x + \frac{h}{2}\right) \cdot \frac{\sin \frac{h}{2}}{\frac{h}{2}} \\
&= \cos x.
\end{aligned}$$

That is
$$(\sin x)' = \cos x.$$

Similarly, we can get
$$(\cos x)' = -\sin x.$$

Example 7 Find the derivative of $f(x) = a^x$ ($a > 0, a \neq 1$).

Solution
$$\begin{aligned}
f'(x) &= \lim_{h \to 0} \frac{f(x+h) - f(x)}{h} \\
&= \lim_{h \to 0} \frac{a^{x+h} - a^x}{h} \\
&= a^x \lim_{h \to 0} \frac{a^h - 1}{h} \\
&\xlongequal{\text{Let } a^h - 1 = t} a^x \lim_{t \to 0} \frac{t}{\log_a(1+t)} \\
&= a^x \frac{1}{\log_a e} = a^x \ln a.
\end{aligned}$$

Specially, we have
$$(e^x)' = e^x.$$

Example 8 Find the derivative of $f(x) = \log_a x$ ($a > 0, a \neq 1$).

Solution
$$\begin{aligned}
f'(x) &= \lim_{h \to 0} \frac{f(x+h) - f(x)}{h} \\
&= \lim_{h \to 0} \frac{\log_a(x+h) - \log_a x}{h}
\end{aligned}$$

$$= \lim_{h\to 0}\frac{1}{h}\log_a\frac{x+h}{x}$$

$$= \frac{1}{x}\lim_{h\to 0}\frac{x}{h}\log_a\left(1+\frac{h}{x}\right)$$

$$= \frac{1}{x}\lim_{h\to 0}\log_a\left(1+\frac{h}{x}\right)^{\frac{x}{h}}$$

$$= \frac{1}{x}\log_a e = \frac{1}{x\ln a},$$

That is
$$(\log_a x)' = \frac{1}{x\ln a}.$$

Specially, we have
$$(\ln x)' = \frac{1}{x}.$$

5. Relationship Between Differentiability and Continuity

Both continuity and differentiability are desirable properties for a function. The following theorem shows how these properties are related:

Theorem 3.1 If the function $f(x)$ is differentiable at a, then $f(x)$ is continuous at a.

Proof Given
$$f'(a) = \lim_{\Delta x\to 0}\frac{\Delta y}{\Delta x} = \lim_{\Delta x\to 0}\frac{f(a+\Delta x)-f(a)}{\Delta x}$$

exists, so
$$\lim_{\Delta x\to 0}\Delta y = \lim_{\Delta x\to 0}\left(\frac{\Delta y}{\Delta x}\cdot \Delta x\right) = \lim_{\Delta x\to 0}\frac{\Delta y}{\Delta x}\cdot \lim_{\Delta x\to 0}\Delta x = 0,$$

then $f(x)$ is continuous at a.

The converse of this theorem is false. If the function $f(x)$ is continuous at a, it does not follow that $f(x)$ has a derivative at a. For example, the function $f(x) = |x|$ is certainly continuous at $x = 0$, but it does not have a derivative there.

Example 9 Discuss the differentiability and continuity of the function
$$f(x) = \begin{cases} x^2 + x, & x \leqslant 1, \\ 2x^3, & x > 1 \end{cases}$$
at $x = 1$.

Solution $x = 1$ is a piecewise point of the function $f(x)$. Since

$$\lim_{x\to 1^+} f(x) = \lim_{x\to 1^+}(x^2+x) = 2,$$
$$\lim_{x\to 1^-} f(x) = \lim_{x\to 1^-} 2x^3 = 2,$$

we can get $\lim_{x\to 1} f(x) = 2 = f(1)$, then $f(x)$ is continuous at $x=1$.

The left-hand derivative of the function $f(x)$ at $x=1$ is
$$f'_-(1) = \lim_{x\to 1^-}\frac{f(x)-f(1)}{x-1} = \lim_{x\to 1^-}\frac{x^2+x-2}{x-1}$$
$$= \lim_{x\to 1^-}\frac{(x-1)(x+2)}{x-1} = \lim_{x\to 1^-}(x+2) = 3.$$

And the right-hand derivative is
$$f'_+(1) = \lim_{x\to 1^+}\frac{f(x)-f(1)}{x-1} = \lim_{x\to 1^+}\frac{2x^3-2}{x-1}$$
$$= 2\lim_{x\to 1^+}\frac{(x-1)(x^2+x+1)}{x-1}$$
$$= 2\lim_{x\to 1^+}(x^2+x+1) = 6.$$

Since the right-hand derivative does not equal to the left-hand derivative at $x=1$, the derivative of the function does not exist.

3.2 The Rules for Finding Derivatives

The process of finding the derivative of a function directly from the definition of the derivative can be time-consuming and tedious. We will develop tools to shortcut this lengthy process, the tools are rules for finding derivatives. We can find derivatives of more complicated functions by the rules.

1. The Constant Multiple Rule

If k is a constant and $f(x)$ is a differentiable function, then
$$[kf(x)]' = kf'(x).$$

The formula says that the derivative of a constant times a function is the constant times the derivative of the function.

Proof Let $F(x)=kf(x)$, then
$$F'(x)=\lim_{h\to 0}\frac{F(x+h)-F(x)}{h}$$
$$=\lim_{h\to 0}\frac{kf(x+h)-kf(x)}{h}$$
$$=\lim_{h\to 0}k\frac{f(x+h)-f(x)}{h}$$
$$=k\lim_{h\to 0}\frac{f(x+h)-f(x)}{h}$$
$$=kf'(x).$$

2. The Sum Rule

If $f(x)$ and $g(x)$ are differentiable functions, then
$$[f(x)+g(x)]'=f'(x)+g'(x).$$

The rule tells us that the derivative of the sum of functions is the sum of the derivatives.

Proof Let $F(x)=f(x)+g(x)$, then
$$F'(x)$$
$$=\lim_{h\to 0}\frac{[f(x+h)+g(x+h)]-[f(x)+g(x)]}{h}$$
$$=\lim_{h\to 0}\left[\frac{f(x+h)-f(x)}{h}+\frac{g(x+h)-g(x)}{h}\right]$$
$$=\lim_{h\to 0}\frac{f(x+h)-f(x)}{h}+\lim_{h\to 0}\frac{g(x+h)-g(x)}{h}$$
$$=f'(x)+g'(x).$$

Remark The Sum Rule can be extended to the sum of any finite number of derivative functions. For instance, using this rule twice, we get
$$(f+g+h)'=[(f+g)+h]'$$
$$=(f+g)'+h'=f'+g'+h'.$$

By writing $f-g$ as $f+(-1)g$, and applying the Sum Rule and the Constant Multiple Rule, we get the following Difference Rule:

3. The Difference Rule

If $f(x)$ and $g(x)$ are differentiable functions, then
$$[f(x)-g(x)]'=f'(x)-g'(x).$$

Example 1 Find the derivatives of the function
$$f(x)=3x^5+x^4-9x^2+5x-1.$$

Solution $f'(x) = (3x^5 + x^4 - 9x^2 + 5x - 1)'$
$= (3x^5)' + (x^4)' - (9x^2)'$
$+ (5x)' - (1)'$
$= 15x^4 + 4x^3 - 18x + 5.$

Next we give the formulas for the derivative of the product and quotient of two functions. So far, we have seen that the derivative of a sum or difference is equal to the sum or difference of the derivatives. We may conjecture that the derivative of a product or quotient is equal to the product or quotient of the derivatives. But this guess is wrong. For example, let $f(x) = x^2$ and $g(x) = x^3$, thus $f'(x) = 2x$, $g'(x) = 3x^2$, $[f(x)g(x)]' = (x^5)' = 5x^4$, but $f'(x)g'(x) \neq [f(x)g(x)]'$. The correct formula was discovered by Leibniz and is called the Product Rule.

4. The Product Rule

If $f(x)$ and $g(x)$ are differentiable functions, then
$$[f(x)g(x)]' = f'(x)g(x) + f(x)g'(x).$$

Example 2 Find the derivative of the function
$$f(x) = (3x^2 - 1)6x^3$$
by applying the Product Rule.

Solution By the Product Rule, we have
$f'(x) = [(3x^2 - 1)6x^3]'$
$= (3x^2 - 1)'6x^3 + (3x^2 - 1)(6x^3)'$
$= 6x \cdot 6x^3 + (3x^2 - 1) \cdot 18x^2$
$= 90x^4 - 18x^2.$

Remark The Product Rule can be also extend to product of any finite number of derivative functions. For instance, using this rule twice, we get
$(fgh)' = (fg)'h + (fg)h'$
$= f'gh + fg'h + fgh'.$

5. The Quotient Rule

If $f(x)$ and $g(x)$ are differentiable functions, and $g(x) \neq 0$, then
$$\left[\frac{f(x)}{g(x)}\right]' = \frac{f'(x)g(x) - f(x)g'(x)}{g^2(x)}.$$

Proof Let $F(x) = \frac{f(x)}{g(x)}$, thus

$$F'(x) = \lim_{h \to 0} \frac{F(x+h) - F(x)}{h}$$
$$= \lim_{h \to 0} \frac{\frac{f(x+h)}{g(x+h)} - \frac{f(x)}{g(x)}}{h}$$
$$= \lim_{h \to 0} \left[\frac{g(x)f(x+h) - f(x)g(x+h)}{h} \cdot \frac{1}{g(x)g(x+h)} \right]$$
$$= \lim_{h \to 0} \left\{ \left[\frac{g(x)f(x+h) - g(x)f(x)}{h} + \frac{g(x)f(x) - f(x)g(x+h)}{h} \right] \cdot \frac{1}{g(x)g(x+h)} \right\}$$
$$= \lim_{h \to 0} \left\{ \left[\frac{f(x+h) - f(x)}{h} g(x) - f(x) \frac{g(x+h) - g(x)}{h} \right] \cdot \frac{1}{g(x)g(x+h)} \right\}$$
$$= [f'(x)g(x) - f(x)g'(x)] \frac{1}{g^2(x)}.$$

Example 3 Find the derivative of the function
$$f(x) = \frac{2x+1}{x^2+1}.$$

Solution
$$f'(x) = \left(\frac{2x+1}{x^2+1} \right)'$$
$$= \frac{(x^2+1)(2x+1)' - (2x+1)(x^2+1)'}{(x^2+1)^2}$$
$$= \frac{2(x^2+1) - (2x+1)2x}{(x^2+1)^2}$$
$$= \frac{-2x^2 - 2x + 2}{(x^2+1)^2}.$$

Example 4 Show that the Power Rule holds for negative exponents, that is
$$(x^{-n})' = -nx^{-n-1} \quad (n > 0).$$

Proof By the Quotient Rule, we have
$$(x^{-n})' = \left(\frac{1}{x^n} \right)' = \frac{0 \cdot x^n - 1 \cdot nx^{n-1}}{x^{2n}}$$
$$= \frac{-nx^{n-1}}{x^{2n}} = -nx^{-n-1}.$$

Example 5 Let the function $y = \tan x$, find y'.

Solution
$$y' = (\tan x)' = \left(\frac{\sin x}{\cos x} \right)'$$

例 3 求函数 $f(x) = \frac{2x+1}{x^2+1}$ 的导数.

解
$$f'(x) = \left(\frac{2x+1}{x^2+1} \right)'$$
$$= \frac{(x^2+1)(2x+1)' - (2x+1)(x^2+1)'}{(x^2+1)^2}$$
$$= \frac{2(x^2+1) - (2x+1)2x}{(x^2+1)^2}$$
$$= \frac{-2x^2 - 2x + 2}{(x^2+1)^2}.$$

例 4 证明：对于负指数的幂法则成立，即
$$(x^{-n})' = -nx^{-n-1} \quad (n > 0).$$

证明 由商法则有
$$(x^{-n})' = \left(\frac{1}{x^n} \right)' = \frac{0 \cdot x^n - 1 \cdot nx^{n-1}}{x^{2n}}$$
$$= \frac{-nx^{n-1}}{x^{2n}} = -nx^{-n-1}.$$

例 5 设函数 $y = \tan x$，求 y'.

解
$$y' = (\tan x)' = \left(\frac{\sin x}{\cos x} \right)'$$

$$= \frac{(\sin x)'\cos x - \sin x(\cos x)'}{\cos^2 x}$$
$$= \frac{\cos^2 x + \sin^2 x}{\cos^2 x} = \frac{1}{\cos^2 x} = \sec^2 x.$$

That is
$$(\tan x)' = \sec^2 x.$$

Example 6 Let the function $y = \sec x$, find y'.

Solution
$$y' = (\sec x)' = \left(\frac{1}{\cos x}\right)'$$
$$= \frac{(1)'\cos x - 1 \cdot (\cos x)'}{\cos^2 x}$$
$$= \frac{\sin x}{\cos^2 x} = \sec x \tan x.$$

That is
$$(\sec x)' = \sec x \tan x.$$

Similarly, We can find the following derivative formulas of $\cot x$ and $\csc x$:
$$(\cot x)' = -\csc^2 x,$$
$$(\csc x) = -\csc x \cot x.$$

6. The Rule for the Derivative of an Inverse Function

Theorem 3.2 If the function $x = f(y)$ is monotonous, differentiable on the interval I_y, and $f'(y) \neq 0$, then the inverse function $y = f^{-1}(x)$ is also differentiable on corresponding interval $I_x = \{x \mid x = f(y), y \in I_y\}$, and
$$[f^{-1}(x)]' = \frac{1}{f'(y)}.$$

This conclusion can be simply stated as the derivatives of inverse function equals to the reciprocal of the derivative of the direct function.

Example 7 Let $x = \sin y \left(y \in \left(-\frac{\pi}{2}, \frac{\pi}{2}\right)\right)$ is the direct function, then $y = \arcsin x$ is its inverse function. The function $x = \sin y$ is monotonous and differentiable on the open interval $\left(-\frac{\pi}{2}, \frac{\pi}{2}\right)$, and
$$(\sin y)' = \cos y > 0.$$

Then, by rule for the derivative of an inverse function, on corresponding interval $I_x = (-1, 1)$, we have

$$(\arcsin x)' = \frac{1}{(\sin y)'} = \frac{1}{\cos y}$$
$$= \frac{1}{\sqrt{1-\sin^2 y}} = \frac{1}{\sqrt{1-x^2}}.$$

That is
$$(\arcsin x)' = \frac{1}{\sqrt{1-x^2}}.$$

Similarly, we have
$$(\arccos x)' = -\frac{1}{\sqrt{1-x^2}}.$$

Example 8 Let $x = \tan y$ $\left(y \in \left(-\frac{\pi}{2}, \frac{\pi}{2}\right)\right)$ is the direct function, then $y = \arctan x$ is the inverse function. The function $x = \tan y$ is monotonous and differentiable on the open interval $\left(-\frac{\pi}{2}, \frac{\pi}{2}\right)$, and
$$(\tan y)' = \sec^2 y \neq 0.$$

Then, by rule for the derivative of an inverse function, on corresponding interval $I_x = (-\infty, +\infty)$, we have
$$(\arctan x)' = \frac{1}{(\tan y)'} = \frac{1}{\sec^2 y}$$
$$= \frac{1}{1+\tan^2 y} = \frac{1}{1+x^2}.$$

That is
$$(\arctan x)' = \frac{1}{1+x^2}.$$

Similarly, we have
$$(\text{arccot}\,x)' = -\frac{1}{1+x^2}.$$

Example 9 Let $x = a^y$ $(a>0, a \neq 1)$ is the direct function, then $y = \log_a x$ is the inverse function. The function $x = a^y$ is monotonous and differentiable on the interval $I_y = (-\infty, +\infty)$, and
$$(a^y)' = a^y \ln a \neq 0.$$

Then, by rule for the derivative of an inverse function, on corresponding interval $I_x = (0, +\infty)$, we have
$$(\log_a x)' = \frac{1}{(a^y)'} = \frac{1}{a^y \ln a} = \frac{1}{x \ln a}.$$

We can see that the result is same to Example 8 in the Section 3.1.

7. The Derivative Formulas of Basic Elementary Functions

(1) $(C)' = 0$ (C is a constant);

(2) $(x^\mu)' = \mu x^{\mu-1}$;

(3) $(\sin x)' = \cos x$;

(4) $(\cos x)' = -\sin x$;

(5) $(\tan x)' = \sec^2 x$;

(6) $(\cot x)' = -\csc^2 x$;

(7) $(\sec x)' = \sec x \tan x$;

(8) $(\csc x)' = -\csc x \cot x$;

(9) $(a^x)' = a^x \ln a$ ($a>0, a \neq 1$);

(10) $(e^x)' = e^x$;

(11) $(\log_a x)' = \dfrac{1}{x \ln a}$ ($a>0, a \neq 1$);

(12) $(\ln x)' = \dfrac{1}{x}$;

(13) $(\arcsin x)' = \dfrac{1}{\sqrt{1-x^2}}$;

(14) $(\arccos x)' = -\dfrac{1}{\sqrt{1-x^2}}$;

(15) $(\arctan x)' = \dfrac{1}{1+x^2}$;

(16) $(\text{arccot}\, x)' = -\dfrac{1}{1+x^2}$.

8. The Chain Rule

Suppose that we are trying to find the derivative of
$$F(x) = (x^2 - x + 4)^{\frac{3}{2}}.$$
If we could let $f(u) = u^{\frac{3}{2}}$, $u = g(x) = x^2 - x + 4$, then $F = f \circ g$. So it would be necessary to have a rule that tells us how to find the derivative of $F = f \circ g$ in terms of the derivatives of f and g. It turns out that the derivative of the composite function $f \circ g$ is the product of the derivatives of f and g. This fact is the most important one of the rules for finding derivatives and is called the **Chain Rule**.

The Chain Rule Let $y = f(u)$ and $u = g(x)$. If $g(x)$ is differentiable at the point x and $f(u)$ is differentiable at the point $u = g(x)$, then the composite function $(f \circ g)(x) = f(g(x))$ is differentiable at x, and

7. 基本初等函数的导数公式

(1) $(C)' = 0$ (C 是常数);

(2) $(x^\mu)' = \mu x^{\mu-1}$;

(3) $(\sin x)' = \cos x$;

(4) $(\cos x)' = -\sin x$;

(5) $(\tan x)' = \sec^2 x$;

(6) $(\cot x)' = -\csc^2 x$;

(7) $(\sec x)' = \sec x \tan x$;

(8) $(\csc x)' = -\csc x \cot x$;

(9) $(a^x)' = a^x \ln a$ ($a>0, a \neq 1$);

(10) $(e^x)' = e^x$;

(11) $(\log_a x)' = \dfrac{1}{x \ln a}$ ($a>0, a \neq 1$);

(12) $(\ln x)' = \dfrac{1}{x}$;

(13) $(\arcsin x)' = \dfrac{1}{\sqrt{1-x^2}}$;

(14) $(\arccos x)' = -\dfrac{1}{\sqrt{1-x^2}}$;

(15) $(\arctan x)' = \dfrac{1}{1+x^2}$;

(16) $(\text{arccot}\, x)' = -\dfrac{1}{1+x^2}$.

8. 链式法则

假设我们要求函数
$$F(x) = (x^2 - x + 4)^{\frac{3}{2}}$$
的导数. 如果令 $f(u) = u^{\frac{3}{2}}$, $u = g(x) = x^2 - x + 4$, 则 $F = f \circ g$. 所以, 有一个告诉我们如何利用 f 和 g 的导数求 $F = f \circ g$ 的导数的法则是必要的. 结果发现, 复合函数 $f \circ g$ 的导数是 f 和 g 的导数的乘积. 这是一个非常重要的求导法则, 我们称之为**链式法则**.

链式法则 设 $y = f(u)$ 和 $u = g(x)$. 如果 $g(x)$ 在点 x 处可导, $f(u)$ 在点 $u = g(x)$ 处可导, 则复合函数 $(f \circ g)(x) = f(g(x))$ 在点 x 处是可导的, 且

$$(f \circ g)'(x) = [f(g(x))]' = f'(u)g'(x)$$
$$= f'(g(x))g'(x),$$

or
$$\frac{dy}{dx} = \frac{dy}{du} \cdot \frac{du}{dx}.$$

We can remember the Chain Rule by this way: The derivative of a composite function is the derivative of the outer function evaluated at the inner function, times the derivative of the inner function.

Example 10 Let $F(x) = (x^2 - x + 4)^{\frac{3}{2}}$, find $F'(x)$.

Solution Here $F(x) = (f \circ g)(x) = f(g(x))$, where $f(u) = u^{\frac{3}{2}}$ and $u = g(x) = x^2 - x + 4$. Since
$$f'(u) = \frac{3}{2} u^{\frac{1}{2}} \quad \text{and} \quad g'(x) = 2x - 1,$$

we have
$$F'(x) = f'(u)g'(x)$$
$$= \frac{3}{2}(x^2 - x + 4)^{\frac{1}{2}}(2x - 1).$$

Example 11 Let $y = \dfrac{1}{(2x^5 - 7)^3}$, find $\dfrac{dy}{dx}$.

Solution Taking $y = \dfrac{1}{u^3} = u^{-3}$, $u = 2x^5 - 7$. By applying the Chain Rule, we have
$$\frac{dy}{dx} = \frac{dy}{du} \cdot \frac{du}{dx} = (u^{-3})'(2x^5 - 7)'$$
$$= (-3u^{-4})(10x^4) = \frac{-30x^4}{(2x^5 - 7)^4}.$$

Example 12 Let $y = \left(\dfrac{x^3 - 2x + 1}{x^4 + 3}\right)^9$, find $\dfrac{dy}{dx}$.

Solution Taking $y = u^9$, $u = \dfrac{x^3 - 2x + 1}{x^4 + 3}$. By applying the Chain Rule, we have
$$\frac{dy}{dx} = \frac{dy}{du} \cdot \frac{du}{dx}$$
$$= 9\left(\frac{x^3 - 2x + 1}{x^4 + 3}\right)^8 \left(\frac{x^3 - 2x + 1}{x^4 + 3}\right)'$$
$$= 9\left(\frac{x^3 - 2x + 1}{x^4 + 3}\right)^8$$
$$\cdot \frac{(3x^2 - 2)(x^4 + 3) - 4x^3(x^3 - 2x + 1)}{(x^4 + 3)^2}$$
$$= 9\left(\frac{x^3 - 2x + 1}{x^4 + 3}\right)^8 \frac{-x^6 + 6x^4 - 4x^3 + 9x^2 - 6}{(x^4 + 3)^2}.$$

$$(f \circ g)'(x) = [f(g(x))]' = f'(u)g'(x)$$
$$= f'(g(x))g'(x)$$

或
$$\frac{dy}{dx} = \frac{dy}{du} \cdot \frac{du}{dx}.$$

可以按这种方式记忆链式法则：复合函数的导数就是外函数在内函数处的导数乘上内函数的导数。

例 10 设 $F(x) = (x^2 - x + 4)^{\frac{3}{2}}$，求 $F'(x)$.

解 这里 $F(x) = (f \circ g)(x) = f(g(x))$，其中 $f(u) = u^{\frac{3}{2}}$，$u = g(x) = x^2 - x + 4$. 因为
$$f'(u) = \frac{3}{2} u^{\frac{1}{2}} \quad \text{和} \quad g'(x) = 2x - 1,$$

我们有
$$F'(x) = f'(u)g'(x)$$
$$= \frac{3}{2}(x^2 - x + 4)^{\frac{1}{2}}(2x - 1).$$

例 11 设 $y = \dfrac{1}{(2x^5 - 7)^3}$，求 $\dfrac{dy}{dx}$.

解 取 $y = \dfrac{1}{u^3} = u^{-3}$，$u = 2x^5 - 7$，则运用链式法则得
$$\frac{dy}{dx} = \frac{dy}{du} \cdot \frac{du}{dx} = (u^{-3})'(2x^5 - 7)'$$
$$= (-3u^{-4})(10x^4) = \frac{-30x^4}{(2x^5 - 7)^4}.$$

例 12 设 $y = \left(\dfrac{x^3 - 2x + 1}{x^4 + 3}\right)^9$，求 $\dfrac{dy}{dx}$.

解 取 $y = u^9$，$u = \dfrac{x^3 - 2x + 1}{x^4 + 3}$，则运用链式法则得
$$\frac{dy}{dx} = \frac{dy}{du} \cdot \frac{du}{dx}$$
$$= 9\left(\frac{x^3 - 2x + 1}{x^4 + 3}\right)^8 \left(\frac{x^3 - 2x + 1}{x^4 + 3}\right)'$$
$$= 9\left(\frac{x^3 - 2x + 1}{x^4 + 3}\right)^8$$
$$\cdot \frac{(3x^2 - 2)(x^4 + 3) - 4x^3(x^3 - 2x + 1)}{(x^4 + 3)^2}$$
$$= 9\left(\frac{x^3 - 2x + 1}{x^4 + 3}\right)^8 \frac{-x^6 + 6x^4 - 4x^3 + 9x^2 - 6}{(x^4 + 3)^2}.$$

Example 13 Let $y=(2x+1)^5(x^3-x+1)^4$, find y'.

Solution We use the Product Rule before using the Chain Rule:
$$y'=(2x+1)^5[(x^3-x+1)^4]'$$
$$+(x^3-x+1)^4[(2x+1)^5]'$$
$$=(2x+1)^5 \cdot 4(x^3-x+1)^3(x^3-x+1)'$$
$$+(x^3-x+1)^4 \cdot 5(2x+1)^4(2x+1)'$$
$$=4(2x+1)^5(x^3-x+1)^3(3x^2-1)$$
$$+5(x^3-x+1)^4(2x+1)^4 \cdot 2$$
$$=2(2x+1)^4(x^3-x+1)^3$$
$$\cdot (17x^3+6x^2-9x+3).$$

The Chain Rule can be generalized to multiple intermediate variables. For instance, let $y=f(u)$, $u=\varphi(v)$, $v=\psi(x)$, then
$$y'=f'(u)\varphi'(v)\psi'(x),$$
or
$$\frac{dy}{dx}=\frac{dy}{du} \cdot \frac{du}{dv} \cdot \frac{dv}{dx}.$$

Example 14 Suppose that $y=\ln\cos e^x$, find y'.

Solution $y'=(\ln\cos e^x)'=\dfrac{1}{\cos e^x}(\cos e^x)'$
$$=\frac{1}{\cos e^x}(-\sin e^x)(e^x)'=-e^x\tan e^x.$$

Example 15 Let $y=e^{\sin\frac{1}{x}}$, find y'.

Solution $y'=\left(e^{\sin\frac{1}{x}}\right)'=e^{\sin\frac{1}{x}}\left(\sin\dfrac{1}{x}\right)'$
$$=e^{\sin\frac{1}{x}}\cos\frac{1}{x} \cdot \left(\frac{1}{x}\right)'$$
$$=-\frac{1}{x^2}e^{\sin\frac{1}{x}}\cos\frac{1}{x}.$$

3.3 Higher-order Derivatives

Suppose that $y=f(x)$ is a differentiable function on the internal I, the derivative of $f(x)$ is $f'(x)$, which is a new function. If the function $f'(x)$ is still differentiable, we produce still another function, denoted by $f''(x)$ and called the **second derivative** of $f(x)$. It may be differentiable, thereby producing $f'''(x)$, which is called the **third** derivative of $f(x)$, and so on. The fourth derivative is denoted by $f^{(4)}(x)$, the fifth derivative is denoted by $f^{(5)}(x)$ in turn. In general, the derivative of the $(n-1)$th derivative of $f(x)$ is called the nth derivative of $f(x)$. When $n \geqslant 4$, the nth derivative of $y=f(x)$ is denoted by

$$f^{(n)}(x), \quad y^{(n)}, \quad \text{or} \quad \frac{d^n y}{dx^n}.$$

For example, let
$$f(x)=6x^4-5x^3+7x^2+8x-1,$$
then
$$f'(x)=24x^3-15x^2+14x+8,$$
$$f''(x)=72x^2-30x+14,$$
$$f'''(x)=144x-30,$$
$$f^{(4)}(x)=144,$$
$$f^{(5)}(x)=0.$$

Since the derivative of the zero function is zero, so the fifth and above derivative of the function $f(x)=6x^4-5x^3+7x^2+8x-1$ is zero.

In general, the second and above derivative of $f(x)$ is called the **higher-order derivative** of $f(x)$, the $f'(x)$ is called the first derivative simultaneously.

Suppose that $u(x)$ and $v(x)$ exist the nth derivatives on the internal I. We can derive the following properties by the definition of the derivative:
$$(u \pm v)^{(n)} = u^{(n)} \pm v^{(n)},$$
$$(uv)^{(n)} = \sum_{k=0}^{n} C_n^k u^{(n-k)} v^{(k)},$$
where $u^{(0)}=0$, $v^{(0)}=0$.

Example 1 Let $y=\sin 2x$, find y''', $y^{(4)}$, and $y^{(12)}$.
Solution
$$y'=2\cos 2x,$$
$$y''=(2\cos 2x)'=-2^2 \sin 2x,$$
$$y'''=(-2^2 \sin 2x)'=-2^3 \cos 2x,$$
$$y^{(4)}=(-2^3 \cos 2x)'=2^4 \sin 2x.$$

We easily have
$$y^{(12)}=2^{12}\sin 2x.$$

Example 2 Let $y=a^x (a>0, a \neq 0)$, find $y^{(n)}$.
Solution
$$y'=a^x \ln a,$$
$$y''=a^x \ln^2 a,$$
$$y'''=a^x \ln^3 a,$$

数记为 $f^{(4)}$，五阶导数记为 $f^{(5)}$，等等. 一般地，函数 $f(x)$ 的 $n-1$ 阶导数的导数称 $f(x)$ 的 n 阶导数. 当 $n \geqslant 4$ 时，$y=f(x)$ 的 n 阶导数记为

$$f^{(n)}(x), \quad y^{(n)} \quad \text{或} \quad \frac{d^n y}{dx^n}.$$

例如，设
$$f(x)=6x^4-5x^3+7x^2+8x-1,$$
则
$$f'(x)=24x^3-15x^2+14x+8,$$
$$f''(x)=72x^2-30x+14,$$
$$f'''(x)=144x-30,$$
$$f^{(4)}(x)=144,$$
$$f^{(5)}(x)=0.$$

由于零函数的导数是零，所以 $f(x)=6x^4-5x^3+7x^2+8x-1$ 的五阶及五阶以上的导数都是零.

通常将 $f(x)$ 的二阶及二阶以上的导数称为**高阶导数**，而相应地将 $f'(x)$ 称为一阶导数.

设 $u(x)$ 和 $v(x)$ 都在区间 I 上存在 n 阶导数. 由导数的定义容易推出以下性质：
$$(u \pm v)^{(n)} = u^{(n)} \pm v^{(n)},$$
$$(uv)^{(n)} = \sum_{k=0}^{n} C_n^k u^{(n-k)} v^{(k)},$$
其中 $u^{(0)}=0$，$v^{(0)}=0$.

例 1 设 $y=\sin 2x$，求 y'''，$y^{(4)}$ 和 $y^{(12)}$.
解
$$y'=2\cos 2x,$$
$$y''=(2\cos 2x)'=-2^2 \sin 2x,$$
$$y'''=(-2^2 \sin 2x)'=-2^3 \cos 2x,$$
$$y^{(4)}=(-2^3 \cos 2x)'=2^4 \sin 2x.$$

容易得到
$$y^{(12)}=2^{12}\sin 2x.$$

例 2 设 $y=a^x (a>0, a \neq 1)$，求 $y^{(n)}$.
解
$$y'=a^x \ln a,$$
$$y''=a^x \ln^2 a,$$
$$y'''=a^x \ln^3 a,$$

$$y^{(4)} = a^x \ln^4 a,$$
......
$$y^{(n)} = a^x \ln^n a.$$

Especially, if $a = e$, then
$$(e^x)^{(n)} = e^x \quad (n = 1, 2, \cdots).$$

Example 3 Let $y = \sin x$, find $y^{(n)}$.

Solution $y' = \cos x = \sin\left(x + \dfrac{\pi}{2}\right),$

$y'' = \cos\left(x + \dfrac{\pi}{2}\right) = \sin\left(x + 2 \cdot \dfrac{\pi}{2}\right),$

$y''' = \cos\left(x + 2 \cdot \dfrac{\pi}{2}\right) = \sin\left(x + 3 \cdot \dfrac{\pi}{2}\right),$

$y^{(4)} = \cos\left(x + 3 \cdot \dfrac{\pi}{2}\right) = \sin\left(x + 4 \cdot \dfrac{\pi}{2}\right),$

......

$y^{(n)} = \sin\left(x + n \cdot \dfrac{\pi}{2}\right).$

Example 4 Let $y = \cos x$, find $y^{(n)}$.

Solution $y' = -\sin x = \cos\left(x + \dfrac{\pi}{2}\right),$

$y'' = -\sin\left(x + \dfrac{\pi}{2}\right) = \cos\left(x + 2 \cdot \dfrac{\pi}{2}\right),$

$y''' = -\sin\left(x + 2 \cdot \dfrac{\pi}{2}\right) = \cos\left(x + 3 \cdot \dfrac{\pi}{2}\right),$

$y^{(4)} = -\sin\left(x + 3 \cdot \dfrac{\pi}{2}\right) = \cos\left(x + 4 \cdot \dfrac{\pi}{2}\right),$

......

$y^{(n)} = \cos\left(x + n \cdot \dfrac{\pi}{2}\right).$

Example 5 Let $y = \ln(1+x)$, find $y^{(n)}$.

Solution $y' = \dfrac{1}{1+x},$

$y'' = -\dfrac{1}{(1+x)^2},$

$y''' = \dfrac{1 \times 2}{(1+x)^3},$

$y^{(4)} = -\dfrac{3!}{(1+x)^4},$

......

$y^{(n)} = (-1)^{n-1} \dfrac{(n-1)!}{(1+x)^n}.$

Example 6 Let $y = x^\mu$ (μ is any constant), find $y^{(n)}$.

Solution $y' = \mu x^{\mu-1}$,
$$y'' = \mu(\mu-1)x^{\mu-2},$$
$$y''' = \mu(\mu-1)(\mu-2)x^{\mu-3},$$
$$y^{(4)} = \mu(\mu-1)(\mu-2)(\mu-3)x^{\mu-4},$$
$$\cdots\cdots$$
$$y^{(n)} = \mu(\mu-1)\cdots(\mu-n+1)x^{\mu-n}.$$

When $\mu = n$, we have
$$(x^n)^{(n)} = n(n-1)(n-2)\cdots 3 \times 2 \times 1$$
$$= n!.$$

But
$$(x^n)^{(n+k)} = 0 \quad (k=1,2,\cdots).$$

Example 7 Let $y = x^2 \mathrm{e}^{2x}$, find $y^{(20)}$.

Solution Let $u = \mathrm{e}^{2x}$, $v = x^2$, then
$$u^{(k)} = 2^k \mathrm{e}^{2x} \quad (k=1,2,\cdots,20)$$
$$v' = 2x, \quad v'' = 2,$$
$$v^{(k)} = 0 \quad (k=3,4,\cdots,20).$$

So we have
$$y^{(20)} = (x^2 \mathrm{e}^{2x})^{(20)} = \sum_{k=0}^{20} \mathrm{C}_{20}^{k} (x^2)^{(k)} (\mathrm{e}^{2x})^{(20-k)}$$
$$= 2^{20} \mathrm{e}^{2x} x^2 + 20 \cdot 2^{19} \mathrm{e}^{2x} \cdot 2x$$
$$+ \frac{20 \cdot 19}{2!} \cdot 2^{18} \mathrm{e}^{2x} \cdot 2$$
$$= 2^{20} \mathrm{e}^{2x} (x^2 + 20x + 95).$$

3.4 The Derivatives of Implicit Functions and Functions Determined by Parameter Equations

1. The Derivative of an Implicit Function

The functions that we have met so far can be described by expressing one variable explicitly in terms of another variable, for example,
$$y = \cos 2x,$$
or, in general, $y = f(x)$. In the equation
$$y - x + \cos y = 0$$
however, we cannot solve for y in terms of x.

3.4 The Derivatives of Implicit Functions and Functions Determined by Parameter Equations

We say that the equation defines y as an **implicit function** of x. If we denote this function by $y(x)$, we can write the equation as
$$y(x) - x + \cos y(x) = 0.$$
Even though we do not have a formula for $y(x)$, we can nevertheless get a relation between x, $y(x)$, and $y'(x)$. By differentiating both sides of the equation with respect to x, we get
$$y'(x) - (x)' + (\cos y(x))' = 0',$$
$$y'(x) - 1 - \sin y(x) \cdot y'(x) = 0,$$
$$y'(x) = \frac{1}{1 - \sin y(x)}.$$

The method just illustrated for finding y' without first solving the given equation for y explicitly in terms of x is called the **method of implicit differentiation**. This consists of differentiating both sides of the equation with respect to x and then solving the resulting equation for y'. In the examples and exercises of this section, it is always assumed that the given equation determines y implicitly as a differentiable function of x so that the method of implicit differentiation can be applied.

Example 1 Find y' if $4x^2 y - 3y = x^3 - 1$.

Solution **Method 1** From the equation, we have
$$(4x^2 - 3)y = x^3 - 1,$$
$$y = \frac{x^3 - 1}{4x^2 - 3}.$$
Thus
$$y' = \frac{(x^3 - 1)'(4x^2 - 3) - (x^3 - 1)(4x^2 - 3)'}{(4x^2 - 3)^2}$$
$$= \frac{3x^2(4x^2 - 3) - 8x(x^3 - 1)}{(4x^2 - 3)^2}$$
$$= \frac{4x^4 - 9x^2 + 8x}{(4x^2 - 3)^2}.$$

Method 2 We equate the derivatives of the two sides:
$$(4x^2 y - 3y)' = (x^3 - 1)',$$
$$8xy + 4x^2 y' - 3y' = 3x^2,$$
$$y' = \frac{-8xy + 3x^2}{4x^2 - 3}.$$

我们称该方程定义了 y 是关于 x 的**隐函数**. 如果用 $y(x)$ 表示这个函数, 则我们可以将此方程记为
$$y(x) - x + \cos y(x) = 0.$$
即使没有一个关于 $y(x)$ 的公式, 我们仍然可以得到 x, $y(x)$ 和 $y'(x)$ 之间的关系. 在方程两边关于 x 求导数, 我们得到
$$y'(x) - (x)' + (\cos y(x))' = 0',$$
$$y'(x) - 1 - \sin y(x) \cdot y'(x) = 0,$$
$$y'(x) = \frac{1}{1 - \sin y(x)}.$$

这种仅求 y' 而不用先将给定方程化为关于 x 的显式函数 y 的方法称为**隐函数求导法**. 这包括对方程两边关于 x 求导数, 然后解所得方程求 y'. 在本节的例题和练习中, 通常假定 y 是 x 的可导隐函数, 可以采用隐函数求导法.

例 1 若 $4x^2 y - 3y = x^3 - 1$, 求 y'.

解 **方法 1** 由原方程, 我们有
$$(4x^2 - 3)y = x^3 - 1,$$
$$y = \frac{x^3 - 1}{4x^2 - 3}.$$
所以
$$y' = \frac{(x^3 - 1)'(4x^2 - 3) - (x^3 - 1)(4x^2 - 3)'}{(4x^2 - 3)^2}$$
$$= \frac{3x^2(4x^2 - 3) - 8x(x^3 - 1)}{(4x^2 - 3)^2}$$
$$= \frac{4x^4 - 9x^2 + 8x}{(4x^2 - 3)^2}.$$

方法 2 我们令两边的导数相等:
$$(4x^2 y - 3y)' = (x^3 - 1)',$$
$$8xy + 4x^2 y' - 3y' = 3x^2,$$
$$y' = \frac{-8xy + 3x^2}{4x^2 - 3}.$$

These two answers look different, but if we substitute $y = \frac{x^3-1}{4x^2-3}$ into the expression just obtained for y', and can get

$$y' = \frac{-8xy+3x^2}{4x^2-3} = \frac{-8x\frac{x^3-1}{4x^2-3}+3x^2}{4x^2-3}$$
$$= \frac{4x^4-9x^2+8x}{(4x^2-3)^2}.$$

Example 2 (1) Find y' if $x^2+y^2=25$.

(2) Find the equation of the tangent line of the circle $x^2+y^2=25$ at the point $(3,-4)$.

Solution (1) we have
$$(x^2+y^2)' = (25)'$$
$$2x+2yy' = 0,$$
$$y' = -\frac{x}{y}.$$

(2) At the point $(3,-4)$, we get
$$y' = \frac{3}{4}.$$

So the equation of the tangent line of the circle $x^2+y^2=25$ at the point $(3,-4)$ is
$$y+4 = \frac{3}{4}(x-3).$$

Example 3 (1) Find y' if $y^5+2y-x-3x^7=0$.

(2) Find the equation of the tangent line of the circle $y^5+2y-x-3x^7=0$ at the point $(0,0)$.

(3) At what point of the circle is the tangent line horizontal?

Solution (1) Differentiating both sides of $y^5+2y-x-3x^7=0$ with respect to x, regarding y as a function of x, using the Chain Rule on the term y^5 and the Constant Multiple Rule on the term $2y$, we have
$$(y^5+2y-x-3x^7)' = (0)',$$
$$5y^4y'+2y'-1-21x^6 = 0,$$
$$y' = \frac{1+21x^6}{5y^4+2}.$$

(2) At the point $(0,0)$, we get
$$y' = \frac{1}{2}.$$

这两个答案看起来不同,但是如果我们将 $y = \frac{x^3-1}{4x^2-3}$ 代入所得到的 y' 的表达式中,有

$$y' = \frac{-8xy+3x^2}{4x^2-3} = \frac{-8x\frac{x^3-1}{4x^2-3}+3x^2}{4x^2-3}$$
$$= \frac{4x^4-9x^2+8x}{(4x^2-3)^2}.$$

例 2 (1) 若 $x^2+y^2=25$,求 y';

(2) 求圆 $x^2+y^2=25$ 在点 $(3,-4)$ 处的切线方程.

解 (1) 我们有
$$(x^2+y^2)' = (25)'$$
$$2x+2yy' = 0,$$
$$y' = -\frac{x}{y}.$$

(2) 在点 $(3,-4)$ 处,我们得到
$$y' = \frac{3}{4}.$$

所以圆 $x^2+y^2=25$ 在点 $(3,-4)$ 处的切线方程是
$$y+4 = \frac{3}{4}(x-3).$$

例 3 (1) 若 $y^5+2y-x-3x^7=0$,求 y';

(2) 求圆 $y^5+2y-x-3x^7=0$ 在点 $(0,0)$ 处的切线方程;

(3) 该圆在哪些点处的切线是水平的?

解 (1) 在方程 $y^5+2y-x-3x^7=0$ 两边关于 x 求导数,将 y 看成 x 的函数,对 y^5 利用链式法则,对 $2y$ 利用数乘法则,我们有
$$(y^5+2y-x-3x^7)' = (0)',$$
$$5y^4y'+2y'-1-21x^6 = 0,$$
$$y' = \frac{1+21x^6}{5y^4+2}.$$

(2) 在点 $(0,0)$ 处,我们得到
$$y' = \frac{1}{2}.$$

3.4 The Derivatives of Implicit Functions and Functions Determined by Parameter Equations

So the equation of the tangent line of the circle $y^5+2y-x-3x^7=0$ at the point $(0,0)$ is
$$y=\frac{1}{2}x.$$

(3) The tangent line is horizontal if $y'=0$, but $1+21x^6\neq 0$ and $5y^4+2\neq 0$, so there isn't any point such that the tangent line is horizontal.

Example 4 Find y' if $x^2+5y^3=x+9$.

Solution $(x^2+5y^3)'=(x+9)'$,
$$2x+15y^2y'=1,$$
$$y'=\frac{1-2x}{15y^2}.$$

Example 5 Find y'' if $x-y+\frac{1}{2}\sin y=0$.

Solution Differentiating the equation implicitly with respect to x, we get
$$1-y'+\frac{1}{2}\cos y\cdot y'=0,$$
$$y'=\frac{2}{2-\cos y}.$$

To find y'', we differentiate this expression for y' using the Quotient Rule and remembering that y is a function of x
$$y''=\frac{-2\sin y\cdot y'}{(2-\cos y)^2}=\frac{-4\sin y}{(2-\cos y)^3}.$$

2. The Derivative of a Function Determined by a Parameter Equation

Some functions can be determined by parameter equations
$$\begin{cases}x=\varphi(t),\\ y=\phi(t)\end{cases}(\alpha\leqslant t\leqslant\beta).$$

For example, the parameter equation of circle is
$$\begin{cases}x=R\cos t,\\ y=R\sin t\end{cases}(0\leqslant t\leqslant 2\pi).$$

The function between x and y is established by parameter t. How do we find the derivative $\frac{dy}{dx}$? Let $\varphi(t)$ and $\phi(t)$ be two derivable functions, $\varphi'(t)\neq 0$, and the inverse function $t=\varphi^{-1}(x)$ exists. We can use the Chain Rule and the rule for the derivative of an inverse function to determine $\frac{dy}{dx}$, that is

$$\frac{\mathrm{d}y}{\mathrm{d}x} = \frac{\mathrm{d}y}{\mathrm{d}t} \cdot \frac{\mathrm{d}t}{\mathrm{d}x} = \frac{\mathrm{d}y}{\mathrm{d}t} \cdot \frac{1}{\frac{\mathrm{d}x}{\mathrm{d}t}}$$

$$= \frac{\phi'(t)}{\varphi'(t)}. \qquad (3.1)$$

If $x = \varphi(t)$ and $y = \phi(t)$ have second derivatives, and $\varphi'(t) \neq 0$, so we can derive

$$\frac{\mathrm{d}^2 y}{\mathrm{d}x^2} = \frac{\mathrm{d}}{\mathrm{d}x}\left(\frac{\mathrm{d}y}{\mathrm{d}x}\right) = \frac{\mathrm{d}}{\mathrm{d}t}\left(\frac{\phi'(t)}{\varphi'(t)}\right) \cdot \frac{\mathrm{d}t}{\mathrm{d}x}$$

$$= \frac{\phi''(t)\varphi'(t) - \phi'(t)\varphi''(t)}{[\varphi'(t)]^2} \cdot \frac{1}{\varphi'(t)}$$

$$= \frac{\phi''(t)\varphi'(t) - \phi'(t)\varphi''(t)}{[\varphi'(t)]^3}. \qquad (3.2)$$

Example 6 Determine the first derivative and second derivative for the following functions:

(1) $\begin{cases} x = 2\mathrm{e}^t, \\ y = \mathrm{e}^{-t}; \end{cases}$

(2) $\begin{cases} x = 1 - t^3, \\ y = t - t^3; \end{cases}$

(3) $\begin{cases} x = a\cos t, \\ y = b\sin t. \end{cases}$

Solution Applying the formulas (3.1) and (3.2), we have

(1) $\dfrac{\mathrm{d}y}{\mathrm{d}x} = \dfrac{(\mathrm{e}^{-t})'}{(2\mathrm{e}^t)'} = \dfrac{-\mathrm{e}^{-t}}{2\mathrm{e}^t},$

$\dfrac{\mathrm{d}^2 y}{\mathrm{d}x^2} = \dfrac{\mathrm{e}^{-t} \cdot 2\mathrm{e}^t - (-\mathrm{e}^{-t}) \cdot 2\mathrm{e}^t}{(2\mathrm{e}^t)^3} = \dfrac{1}{2\mathrm{e}^{3t}}.$

(2) $\dfrac{\mathrm{d}y}{\mathrm{d}x} = \dfrac{(t - t^3)'}{(1 - t^3)'} = \dfrac{1 - 3t^2}{-3t^2},$

$\dfrac{\mathrm{d}^2 y}{\mathrm{d}x^2} = \dfrac{(1 - 3t^2)' \cdot (-3t^2) - (1 - 3t^2) \cdot (-3t^2)'}{(-3t^2)^3}$

$= \dfrac{2}{-9t^5}.$

(3) $\dfrac{\mathrm{d}y}{\mathrm{d}x} = \dfrac{(b\sin t)'}{(a\cos t)'} = \dfrac{b\cos t}{-a\sin t},$

$\dfrac{\mathrm{d}^2 y}{\mathrm{d}x^2} = \dfrac{(b\cos t)' \cdot (-a\sin t) - (b\cos t) \cdot (-a\sin t)'}{(-a\sin t)^3}$

$= \dfrac{b(\sin^2 t + \cos^2 t)}{-a^2 \sin^3 t} = -\dfrac{b}{a^2 \sin^3 t}.$

$$\frac{\mathrm{d}y}{\mathrm{d}x} = \frac{\mathrm{d}y}{\mathrm{d}t} \cdot \frac{\mathrm{d}t}{\mathrm{d}x} = \frac{\mathrm{d}y}{\mathrm{d}t} \cdot \frac{1}{\frac{\mathrm{d}x}{\mathrm{d}t}}$$

$$= \frac{\phi'(t)}{\varphi'(t)}. \qquad (3.1)$$

如果 $x = \varphi(t)$ 和 $y = \phi(t)$ 是二阶可导的,且 $\varphi'(t) \neq 0$,所以我们有

$$\frac{\mathrm{d}^2 y}{\mathrm{d}x^2} = \frac{\mathrm{d}}{\mathrm{d}x}\left(\frac{\mathrm{d}y}{\mathrm{d}x}\right) = \frac{\mathrm{d}}{\mathrm{d}t}\left(\frac{\phi'(t)}{\varphi'(t)}\right) \cdot \frac{\mathrm{d}t}{\mathrm{d}x}$$

$$= \frac{\phi''(t)\varphi'(t) - \phi'(t)\varphi''(t)}{[\varphi'(t)]^2} \cdot \frac{1}{\varphi'(t)}$$

$$= \frac{\phi''(t)\varphi'(t) - \phi'(t)\varphi''(t)}{[\varphi'(t)]^3}. \qquad (3.2)$$

例 6 求下列函数的一阶和二阶导数:

(1) $\begin{cases} x = 2\mathrm{e}^t, \\ y = \mathrm{e}^{-t}; \end{cases}$

(2) $\begin{cases} x = 1 - t^3, \\ y = t - t^3; \end{cases}$

(3) $\begin{cases} x = a\cos t, \\ y = b\sin t. \end{cases}$

解 应用公式(3.1)和(3.2),我们有

(1) $\dfrac{\mathrm{d}y}{\mathrm{d}x} = \dfrac{(\mathrm{e}^{-t})'}{(2\mathrm{e}^t)'} = \dfrac{-\mathrm{e}^{-t}}{2\mathrm{e}^t},$

$\dfrac{\mathrm{d}^2 y}{\mathrm{d}x^2} = \dfrac{\mathrm{e}^{-t} \cdot 2\mathrm{e}^t - (-\mathrm{e}^{-t}) \cdot 2\mathrm{e}^t}{(2\mathrm{e}^t)^3} = \dfrac{1}{2\mathrm{e}^{3t}}.$

(2) $\dfrac{\mathrm{d}y}{\mathrm{d}x} = \dfrac{(t - t^3)'}{(1 - t^3)'} = \dfrac{1 - 3t^2}{-3t^2},$

$\dfrac{\mathrm{d}^2 y}{\mathrm{d}x^2} = \dfrac{(1 - 3t^2)' \cdot (-3t^2) - (1 - 3t^2) \cdot (-3t^2)'}{(-3t^2)^3}$

$= \dfrac{2}{-9t^5}.$

(3) $\dfrac{\mathrm{d}y}{\mathrm{d}x} = \dfrac{(b\sin t)'}{(a\cos t)'} = \dfrac{b\cos t}{-a\sin t},$

$\dfrac{\mathrm{d}^2 y}{\mathrm{d}x^2} = \dfrac{(b\cos t)' \cdot (-a\sin t) - (b\cos t) \cdot (-a\sin t)'}{(-a\sin t)^3}$

$= \dfrac{b(\sin^2 t + \cos^2 t)}{-a^2 \sin^3 t} = -\dfrac{b}{a^2 \sin^3 t}.$

3.5 The Differential and the Approximation
3.5 微分和近似

The notation $\dfrac{\mathrm{d}y}{\mathrm{d}x}$ has been used to mean the derivative of y with respect to x. The notation $\dfrac{\mathrm{d}}{\mathrm{d}x}$ has been used to mean the derivative with respect to x. Up to now, we have treated $\dfrac{\mathrm{d}y}{\mathrm{d}x}$ (or $\dfrac{\mathrm{d}}{\mathrm{d}x}$) as a single symbol and have not tried to give separate meanings to the symbols $\mathrm{d}y$ and $\mathrm{d}x$. In this section, we will give meanings of $\mathrm{d}y$ and $\mathrm{d}x$.

Let the function $f(x)$ be differentiable. To motivate our definitions, let $P(x_0, y_0)$ be a point on the curve $y = f(x)$. Since $f(x)$ is differentiable, the following limit exists:
$$f'(x_0) = \lim_{\Delta x \to 0} \frac{f(x_0 + \Delta x) - f(x_0)}{\Delta x}$$

Thus, if Δx is small, the quotient
$$\frac{f(x_0 + \Delta x) - f(x_0)}{\Delta x}$$
will approximate $f'(x_0)$, so we have
$$f(x_0 + \Delta x) - f(x_0) \approx f'(x_0) \Delta x.$$

The left side of this expression is Δy, this is the actual change in y as x changes from x_0 to $x_0 + \Delta x$. The right side is equal to the change of ordinate in the tangent line as x changes from x_0 to $x_0 + \Delta x$, it is denoted $\mathrm{d}y$, and it serves as an approximation to Δy (Figure 3.4). When Δx is small, we hope $\mathrm{d}y$ to be a good approximation to Δy, it is usually easier to calculate. Next we will give the definitions of the differentials $\mathrm{d}y$ and $\mathrm{d}x$.

符号 $\dfrac{\mathrm{d}y}{\mathrm{d}x}$ 常常被用来表示 y 关于 x 的导数，而符号 $\dfrac{\mathrm{d}}{\mathrm{d}x}$ 被用来表示关于 x 的导数. 到目前为止, 我们将 $\dfrac{\mathrm{d}y}{\mathrm{d}x}$ （或 $\dfrac{\mathrm{d}}{\mathrm{d}x}$）作为一个整体符号, 还不曾单独给出符号 $\mathrm{d}y$ 和 $\mathrm{d}x$ 的含义. 在这节中, 我们将要给 $\mathrm{d}y$ 和 $\mathrm{d}x$ 赋予意义.

设 $f(x)$ 是可导函数. 为了给出我们的定义, 令 $P(x_0, y_0)$ 是曲线 $y = f(x)$ 上的点. 由于 $f(x)$ 可导, 所以下面的极限存在:
$$f'(x_0) = \lim_{\Delta x \to 0} \frac{f(x_0 + \Delta x) - f(x_0)}{\Delta x}$$

于是, 如果 Δx 很小, 商
$$\frac{f(x_0 + \Delta x) - f(x_0)}{\Delta x}$$
将会近似于 $f'(x_0)$, 从而我们有
$$f(x_0 + \Delta x) - f(x_0) \approx f'(x_0) \Delta x.$$

这个表达式的左边为 Δy, 它实际上是当 x 由 x_0 变到 $x_0 + \Delta x$ 时 y 的变化量. 表达式的右边是当 x 由 x_0 变动到 $x_0 + \Delta x$ 时切线纵坐标的变化量, 将其记为 $\mathrm{d}y$, 它可以看成 Δy 的近似（图 3.4）. 当 Δx 很小时, 我们希望 $\mathrm{d}y$ 是 Δy 的很好的近似, 也很容易计算. 下面我们来给出 $\mathrm{d}y$ 和 $\mathrm{d}x$ 的定义.

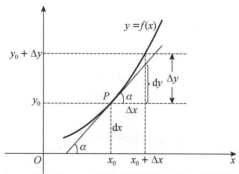

Figure 3.4
图 3.4

1. The Definition of the Differential

Let $y=f(x)$ be a differentiable function, Δx is an arbitrary increment in the independent variable x, Δy is the actual change in the variable y as x changes from x_0 to $x_0+\Delta x$, that is
$$\Delta y=f(x_0+\Delta x)-f(x_0).$$
$\mathrm{d}x$ is called the **differential** of the independent variable x, is defined by
$$\mathrm{d}x=\Delta x.$$
$\mathrm{d}y$ is called the **differential** of the dependent variable y at x_0, is defined by
$$\mathrm{d}y=f'(x_0)\mathrm{d}x.$$

If the function $y=f(x)$ is differentiable on the interval I, $\mathrm{d}y=f'(x)\mathrm{d}x$ is called the **differential** of the function $y=f(x)$.

Example 1 Let $y=f(x)=x^3+x^2-2x+1$, compute the values of Δy and $\mathrm{d}y$, where x changes (1) from 2 to 2.05 and (2) from 2 to 2.01.

Solution (1) We have
$$f(2)=2^3+2^2-2\times 2+1=9,$$
$$f(2.05)=2.05^3+2.05^2-2\times 2.05+1$$
$$=9.717\ 625,$$
$$\Delta y=f(2.05)-f(2)=0.717\ 625.$$
By the definitions of differential, we have
$$\mathrm{d}y=f'(x)\mathrm{d}x=(3x^2+2x-2)\mathrm{d}x.$$
When $x=2$ and $\mathrm{d}x=\Delta x=0.05$, this becomes
$$\mathrm{d}y=(3\times 2^2+2\times 2-2)\times 0.05=0.7.$$
(2) $f(2.01)=2.01^3+2.01^2-2\times 2.01+1$
$$=9.140\ 701,$$
$$\Delta y=f(2.01)-f(2)=0.140\ 701.$$
When $x=2$ and $\mathrm{d}x=\Delta x=0.01$, we have
$$\mathrm{d}y=(3\times 2^2+2\times 2-2)\times 0.01=0.14.$$

Remark The approximation $\Delta y\approx\mathrm{d}y$ becomes better as Δx becomes smaller in above example, and $\mathrm{d}y$ is easier to compute than Δy. For more complicated functions, it may be impossible to compute Δy exactly. In such cases, the approximation by differentials is especially useful.

From the Chain Rule of the derivative of complex function, we can get the following property:

1. 微分的定义

设 $y=f(x)$ 是可导函数，Δx 是一个有关自变量 x 的任意增量，Δy 是当 x 由 x_0 变动 $x_0+\Delta x$ 时变量 y 的实际变化量，即
$$\Delta y=f(x_0+\Delta x)-f(x_0).$$
$\mathrm{d}x$ 称为自变量 x 的**微分**，定义为
$$\mathrm{d}x=\Delta x.$$
$\mathrm{d}y$ 称为因变量 y 在点 x_0 处的**微分**，定义为
$$\mathrm{d}y=f'(x_0)\mathrm{d}x.$$

如果函数 $y=f(x)$ 在区间 I 上可导，则称 $\mathrm{d}y=f'(x)\mathrm{d}x$ 为函数 $y=f(x)$ 的**微分**.

例 1 设 $y=f(x)=x^3+x^2-2x+1$，求 Δy 和 $\mathrm{d}y$ 的值，其中 (1) x 由 2 变到 2.05；(2) x 由 2 变到 2.01.

解 (1) 我们有
$$f(2)=2^3+2^2-2\times 2+1=9,$$
$$f(2.05)=2.05^3+2.05^2-2\times 2.05+1$$
$$=9.717\ 625,$$
$$\Delta y=f(2.05)-f(2)=0.717\ 625.$$
由微分的定义有
$$\mathrm{d}y=f'(x)\mathrm{d}x=(3x^2+2x-2)\mathrm{d}x.$$
当 $x=2$ 和 $\mathrm{d}x=\Delta x=0.05$ 时，有
$$\mathrm{d}y=(3\times 2^2+2\times 2-2)\times 0.05=0.7.$$
(2) $f(2.01)=2.01^3+2.01^2-2\times 2.01+1$
$$=9.140\ 701,$$
$$\Delta y=f(2.01)-f(2)=0.140\ 701.$$
当 $x=2$ 和 $\mathrm{d}x=\Delta x=0.01$ 时，有
$$\mathrm{d}y=(3\times 2^2+2\times 2-2)\times 0.01=0.14.$$

注 由上例可以看出 Δx 变动得越小，$\Delta y\approx\mathrm{d}y$ 近似得越好，并且 $\mathrm{d}y$ 比 Δy 更容易计算. 对于更复杂的函数，精确地计算 Δy 是不可能的. 在这些例子中，由微分得到的近似更有用.

由复合函数的求导的链式法则可推导出如下性质：

If $y=f(u)$ and $u=g(x)$ are differentiable, the differential of the complex function $y=f(g(x))$ is
$$\mathrm{d}y=y'_x\mathrm{d}x=f'(u)g'(x)\mathrm{d}x.$$
Because of $g(x)\mathrm{d}x=\mathrm{d}u$, the differential of the complex function $y=f(g(x))$ is also
$$\mathrm{d}y=f'(u)\mathrm{d}u.$$
This form is consistent with the form which u is the independent variable. So we call it the **invariance of differential form**.

Example 2 Find the differentials of the following functions：
(1) $y=x^3-3x+1$;
(2) $y=\sin(2x+1)$;
(3) $y=\ln(1+e^{x^2})$.

Solution (1) $\mathrm{d}y=(x^3-3x+1)'\mathrm{d}x$
$\qquad =(3x^2-3)\mathrm{d}x.$

(2) $\mathrm{d}y=\mathrm{d}(\sin(2x+1))$
$\qquad =\cos(2x+1)\mathrm{d}(2x+1)$
$\qquad =2\cos(2x+1)\mathrm{d}x.$

(3) $\mathrm{d}y=\mathrm{d}(\ln(1+e^{x^2}))$
$\qquad =\dfrac{1}{1+e^{x^2}}\mathrm{d}(1+e^{x^2})$
$\qquad =\dfrac{1}{1+e^{x^2}}e^{x^2}\mathrm{d}(x^2)$
$\qquad =\dfrac{2x}{1+e^{x^2}}e^{x^2}\mathrm{d}x.$

Corresponding to the derivative rules and the derivative formulas of basic elementary functions by multiplying $\mathrm{d}x$, we will get the rules of differential and the differential formulas of basic elementary functions in the following：

2. The Rules of the Differential

(1) $\mathrm{d}(u\pm v)=\mathrm{d}u\pm \mathrm{d}v$;
(2) $\mathrm{d}(ku)=k\mathrm{d}u$ (k is a constant);
(3) $\mathrm{d}(uv)=v\mathrm{d}u+u\mathrm{d}v$;
(4) $\mathrm{d}\left(\dfrac{u}{v}\right)=\dfrac{v\mathrm{d}u-u\mathrm{d}v}{v^2}$ ($v\neq 0$).

Note that $u=u(x)$ and $v=v(x)$ are two differentiable functions.

如果 $y=f(u)$ 和 $u=g(x)$ 都可导，则复合函数 $y=f(g(x))$ 的微分为
$$\mathrm{d}y=y'_x\mathrm{d}x=f'(u)g'(x)\mathrm{d}x.$$
由于 $g(x)\mathrm{d}x=\mathrm{d}u$，复合函数 $y=f(g(x))$ 的微分也可以写为
$$\mathrm{d}y=f'(u)\mathrm{d}u.$$
这一形式与 u 为自变量时的微分形式一致，所以我们称其为**微分形式的不变性**.

例 2 求下列函数的微分：
(1) $y=x^3-3x+1$;
(2) $y=\sin(2x+1)$;
(3) $y=\ln(1+e^{x^2})$.

解 (1) $\mathrm{d}y=(x^3-3x+1)'\mathrm{d}x$
$\qquad =(3x^2-3)\mathrm{d}x.$

(2) $\mathrm{d}y=\mathrm{d}(\sin(2x+1))$
$\qquad =\cos(2x+1)\mathrm{d}(2x+1)$
$\qquad =2\cos(2x+1)\mathrm{d}x.$

(3) $\mathrm{d}y=\mathrm{d}(\ln(1+e^{x^2}))$
$\qquad =\dfrac{1}{1+e^{x^2}}\mathrm{d}(1+e^{x^2})$
$\qquad =\dfrac{1}{1+e^{x^2}}e^{x^2}\mathrm{d}(x^2)$
$\qquad =\dfrac{2x}{1+e^{x^2}}e^{x^2}\mathrm{d}x.$

相应于求导法则和基本初等函数的导数公式，通过乘以 $\mathrm{d}x$ 就可以得到下面的微分法则和基本初等函数的微分公式：

2. 微分法则

(1) $\mathrm{d}(u\pm v)=\mathrm{d}u\pm \mathrm{d}v$;
(2) $\mathrm{d}(ku)=k\mathrm{d}u$ (k 是常数);
(3) $\mathrm{d}(uv)=v\mathrm{d}u+u\mathrm{d}v$;
(4) $\mathrm{d}\left(\dfrac{u}{v}\right)=\dfrac{v\mathrm{d}u-u\mathrm{d}v}{v^2}$ ($v\neq 0$).

注意，这里 $u=u(x)$，$v=v(x)$ 是两个可导函数.

3. **The Differential Formulas of Basic Elementary Functions**

(1) $d(C) = 0$ (C is a constant);

(2) $d(x^\mu) = \mu x^{\mu-1} dx$ ($\mu \neq 1$);

(3) $d(\sin x) = \cos x \, dx$;

(4) $d(\cos x) = -\sin x \, dx$;

(5) $d(\tan x) = \sec^2 x \, dx$;

(6) $d(\cot x) = -\csc^2 x \, dx$;

(7) $d(\sec x) = \sec x \tan x \, dx$;

(8) $d(\csc x) = -\csc x \cot x \, dx$;

(9) $d(a^x) = a^x \ln a \, dx$ ($a > 0$, $a \neq 1$);

(10) $d(e^x) = e^x dx$;

(11) $d(\log_a x) = \dfrac{1}{x \ln a} dx$ ($a > 0$, $a \neq 1$);

(12) $d(\ln x) = \dfrac{1}{x} dx$.

4. **The Linear Approximation of a Function**

If the function $f(x)$ is differentiable at a, then from the point-slope form of a line, the tangent line of the curve $y = f(x)$ at $(a, f(a))$ is given by
$$y = f(a) + f'(a)(x - a).$$
The function
$$L(x) = f(a) + f'(a)(x - a)$$
is called the **linear approximation** of the function $f(x)$ at a, and it is often a very good approximation of $f(x)$ when x is infinitely close to a.

Example 3 Find the linear approximation of $f(x) = 1 + \sin 2x$ at $x = \dfrac{\pi}{2}$.

Solution The derivative of $f(x)$ is
$$f'(x) = 2\cos 2x,$$
so the linear approximation is
$$\begin{aligned} L(x) &= f\left(\dfrac{\pi}{2}\right) + f'\left(\dfrac{\pi}{2}\right)\left(x - \dfrac{\pi}{2}\right) \\ &= 1 + \sin\pi + 2\cos\pi\left(x - \dfrac{\pi}{2}\right) \\ &= 1 - 2\left(x - \dfrac{\pi}{2}\right) \\ &= 1 + \pi - 2x. \end{aligned}$$

3. 基本初等函数的微分公式

(1) $d(C) = 0$ (C 是常数);

(2) $d(x^\mu) = \mu x^{\mu-1} dx$ ($\mu \neq 1$);

(3) $d(\sin x) = \cos x \, dx$;

(4) $d(\cos x) = -\sin x \, dx$;

(5) $d(\tan x) = \sec^2 x \, dx$;

(6) $d(\cot x) = -\csc^2 x \, dx$;

(7) $d(\sec x) = \sec x \tan x \, dx$;

(8) $d(\csc x) = -\csc x \cot x \, dx$;

(9) $d(a^x) = a^x \ln a \, dx$ ($a > 0$, $a \neq 1$);

(10) $d(e^x) = e^x dx$;

(11) $d(\log_a x) = \dfrac{1}{x \ln a} dx$ ($a > 0$, $a \neq 1$);

(12) $d(\ln x) = \dfrac{1}{x} dx$.

4. 函数的线性近似

如果函数 $f(x)$ 在点 a 处可导,则由直线的点斜式方程可知曲线 $y = f(x)$ 在点 $(a, f(a))$ 处的切线为
$$y = f(a) + f'(a)(x - a).$$
函数
$$L(x) = f(a) + f'(a)(x - a)$$
称为函数 $f(x)$ 在点 a 处的**线性近似**. 当 x 趋近 a 时,$L(x)$ 就是 $f(x)$ 的一个很好的线性近似.

例 3 求 $f(x) = 1 + \sin 2x$ 在点 $x = \dfrac{\pi}{2}$ 处的线性近似.

解 $f(x)$ 的导数是
$$f'(x) = 2\cos 2x,$$
所以所求的线性近似为
$$\begin{aligned} L(x) &= f\left(\dfrac{\pi}{2}\right) + f'\left(\dfrac{\pi}{2}\right)\left(x - \dfrac{\pi}{2}\right) \\ &= 1 + \sin\pi + 2\cos\pi\left(x - \dfrac{\pi}{2}\right) \\ &= 1 - 2\left(x - \dfrac{\pi}{2}\right) \\ &= 1 + \pi - 2x. \end{aligned}$$

Exercises 3
习题 3

1. Find the derivatives of the following functions:

(1) $y = x^2 + x + 1$;

(2) $y = \dfrac{3}{x^2+1}$;

(3) $y = \dfrac{x+2}{x-2}$;

(4) $y = \sqrt{2x}$;

(5) $y = (x^4 - 2x)(x^2 - 3x + 1)$;

(6) $y = \dfrac{2}{x}$;

(7) $y = \dfrac{100}{x^3}$;

(8) $y = 3\cos x + 2\sin x$;

(9) $y = x\cot x + \csc x$;

(10) $y = \dfrac{2\sec x}{1+\sec x}$;

(11) $y = -\dfrac{\cos x}{x^2}$;

(12) $y = \dfrac{x\sin x + \cos x}{x^2 + 1}$.

2. If $f(0) = 7$, $f'(0) = 2$, $g(0) = 6$, and $g'(0) = -10$, find $(fg)'(0)$, $(f+g)'(0)$, and $\left(\dfrac{f}{g}\right)'(0)$.

3. Find the equation of the tangent line of $f(x) = \dfrac{1}{x}$ at the point $\left(\dfrac{1}{2}, 2\right)$.

4. Find the equation of the tangent line of $f(x) = x^{3/2}$ at the point $(0, -4)$.

5. Find the derivatives of the following composite functions:

(1) $y = \left(\dfrac{2x-1}{3x+2}\right)^3$;

(2) $y = (3-2x)^3$;

(3) $y = \cos(x+3)^2$;

(4) $y = \left(\dfrac{x^2}{1+\sqrt{x}}\right)^2$;

(5) $y=(x^6+3)^2(2-x)^4$;

(6) $y=\dfrac{(3x-11)^3}{2x+5}$;

(7) $y=e^x(x^2+2x-3)^2$;

(8) $y=\ln\sin\dfrac{1}{x}$;

(9) $y=\sin\left(\cos\dfrac{1}{x^2}\right)$;

(10) $y=\sin^3 2x-3\ln(\pi x^4)$.

6. Find the higher derivatives of the following functions:

(1) $y=ax+b$, find y'';

(2) $s=\sin\omega t$, find s'';

(3) $y=\dfrac{1}{x}$, find $y^{(n)}$.

7. Check if the function $y=\sqrt{2x-x^2}$ meets
$$y^3 y''+1=0.$$

8. Find $\dfrac{dy}{dx}$ by implicit differentiation:

(1) $y^2-2xy+9=0$;

(2) $x^3+y^3-4\pi xy=0$;

(3) $xy=1$;

(4) $xy^2=x-1$;

(5) $\tan\dfrac{x}{y}=2y-x$;

(6) $2\cos x\sin 2y=5$;

(7) $yx=\cot x^2+\csc xy$;

(8) $2\sqrt{xy}=xy^2+2$;

(9) $x^2+y^2=2y$;

(10) $y^3-xy+2\cos xy=2$.

9. If $y+x^2 y^3=10$ and $y(1)=2$, find $y'(1)$.

10. If $y^5+2y-x-3x^7=0$, find $\dfrac{d^2 y}{dx^2}$.

11. If $x^2+y^2=25$, find $\dfrac{d^3 y}{dx^3}$.

12. If $x^3-4y^2+3=0$, find $\dfrac{d^3 y}{dx^3}$.

13. If $xy+y^3=2$, find $\dfrac{d^3 y}{dx^3}$.

14. Find the differentials of the following functions:

(1) $y = 3x^2$;

(2) $y = \dfrac{1}{x} + 2\sqrt{x}$;

(3) $y = x^3 - 2x + 1$;

(4) $y = (x^7 - 5x + 3)^2$;

(5) $y = 2x^4 + 3x^3 + 4x^2 + 5x + 6$;

(6) $y = 2\sin x(x^{-3} + \pi x^{-5})$;

(7) $y = \left(\dfrac{2}{3}x^3 + 2\right)(x+3)^2$;

(8) $y = \dfrac{2}{x} - \dfrac{1}{x^2}$;

(9) $y = 2x(x^{10} + \sqrt{\cos 2x})$;

(10) $y = \dfrac{3x+2}{2x-1}$;

(11) $y = (x^4 - 2\cos x)(x^2 - 3x + 1)$.

15. If $y = f(x) = x^4 - 2x + 1$, compute the values of Δy and dy, where x changes (1) from 3 to 3.01 and (2) from 2 to 2.05.

16. If $y = f(x) = \sqrt{4 + 5x}$, compute the values of Δy and dy, where x changes (1) from 0 to 0.01 and (2) from 1 to 1.05.

17. Find the linear approximation of $f(x) = \tan 2x$ at $x = \pi/8$.

18. Find the linear approximation of $f(x) = x^2 \sin 2x$ at $x = \pi/6$.

14. 计算下列函数的微分：

(1) $y = 3x^2$；

(2) $y = \dfrac{1}{x} + 2\sqrt{x}$；

(3) $y = x^3 - 2x + 1$；

(4) $y = (x^7 - 5x + 3)^2$；

(5) $y = 2x^4 + 3x^3 + 4x^2 + 5x + 6$；

(6) $y = 2\sin x(x^{-3} + \pi x^{-5})$；

(7) $y = \left(\dfrac{2}{3}x^3 + 2\right)(x+3)^2$；

(8) $y = \dfrac{2}{x} - \dfrac{1}{x^2}$；

(9) $y = 2x(x^{10} + \sqrt{\cos 2x})$；

(10) $y = \dfrac{3x+2}{2x-1}$；

(11) $y = (x^4 - 2\cos x)(x^2 - 3x + 1)$.

15. 若 $y = f(x) = x^4 - 2x + 1$，计算 Δy 和 dy 的值，其中(1) x 由 3 变到 3.01；(2) x 由 2 变到 2.05.

16. 若 $y = f(x) = \sqrt{4 + 5x}$，计算 Δy 和 dy 的值，其中(1) x 由 0 变到 0.01；(2) x 由 1 变到 1.05.

17. 求 $f(x) = \tan 2x$ 在 $x = \pi/8$ 处的线性近似.

18. 求 $f(x) = x^2 \sin 2x$ 在 $x = \pi/6$ 处的线性近似.

Chapter 4 Applications of the Derivative
第 4 章 导数的应用

In the previous chapter, we began with the change speed degree of the dependent variable relative to independent variable form analysis of practical problems, introduced the concept of derivative, and discussed the calculation method of derivative. In this chapter, we will use derivatives to investigate some properties of functions and curves, and use these knowledge to solve some practical problems. Therefore, firstly, we introduce some mean value theorems of differentiation calculus (Rolle Theorem, Lagrange Mean Value Theorem and Cauchy Mean Value Theorem), they are the theoretical basis of the application of derivatives.

在上一章中,我们从分析实际问题中因变量相对于自变量的变化快慢程度出发,引进了导数的概念,并讨论了导数的计算方法. 在这一章中,我们将应用导数来研究函数以及曲线的某些性态,并利用这些知识解决一些实际问题. 为此,先要介绍微分学中的几个中值定理(罗尔定理、拉格朗日中值定理和柯西中值定理),它们是导数应用的理论基础.

4.1 The Mean Value Theorem
4.1 微分中值定理

The Fermat's Lemma Suppose that $f(x)$ is defined on an open interval containing x_0. If $f(x_0)$ is maximum value or minimum value on the interval, and $f'(x_0)$ exists, then
$$f'(x_0) = 0.$$

费马引理 设 $f(x)$ 在包含 x_0 的某开区间内有定义. 如果 $f(x_0)$ 在这个区间上取得最大值或者最小值,并且 $f'(x_0)$ 存在,那么必有
$$f'(x_0) = 0.$$

The Rolle Theorem Let $f(x)$ be a function that satisfies the following three hypotheses:
(1) $f(x)$ is continuous on the closed interval $[a,b]$;
(2) $f(x)$ is differentiable on the open interval (a,b);
(3) $f(a) = f(b)$.
Then, there is at least one number c in (a,b) such that
$$f'(c) = 0.$$
The geometric meaning of the Rolle Theorem: If the curve AB is the graph of the function $f(x)$, it is a continuous

罗尔定理 设函数 $f(x)$ 满足下面三个条件:
(1) 在闭区间 $[a,b]$ 上连续;
(2) 在开区间 (a,b) 内可导;
(3) $f(a) = f(b)$,
则在区间 (a,b) 内至少存在一点 c,使得
$$f'(c) = 0.$$
罗尔定理的几何意义是:设曲线弧 AB 是函数 $y = f(x)$ 的图像,它是一条连续的曲

curve and has a non vertical tangent line at every point between A and B, and the heights of two end points of the curve are equal, then there is at least one point C on the graph between A and B at which the tangent line is parallel (Figure 4.1).

线弧,除端点外处处有不垂直于 x 轴的切线,且两个端点的高度相等,则在曲线弧 AB 上至少存在一点 C,使得在 C 处曲线有水平的切线(图 4.1).

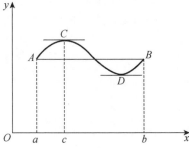

Figure 4.1
图 4.1

Proof Since $f(x)$ is continous on the closed interval $[a,b]$, by the Max-Min Existence Theorem, $f(x)$ has a maximum value M and a minimum value m somewhere in $[a,b]$. There are two conditions:

Case 1 $M=m$. Then $f(x)$ gets same values on $[a,b]$, that is $f(x)=M$, a constant. So, for any $x\in(a,b)$, we have $f'(x)=0$. The number c can be taken to be any number in (a,b).

Case 2 $M>m$. Since $f(a)=f(b)$, it must attain its maximum value M or minimum value m at a number c in the open interval (a,b). For definitude, we assume that it must attain its maximum value at a number c in the open interval (a,b). By hypothesis (2), $f(x)$ is differentiable at c. Thus, $f'(c)=0$ by the Fermat's Lemma.

The Lagrange Mean Value Theorem Let $f(x)$ be a function that fulfills two hypotheses:

(1) $f(x)$ is continuous on the closed interval $[a,b]$;

(2) $f(x)$ is differentiable on the open interval (a,b).

Then, there is at least one number c in (a,b) such that

$$\frac{f(b)-f(a)}{b-a}=f'(c)$$

or, equivalently,

$$f(b)-f(a)=f'(c)(b-a). \tag{4.1}$$

证明 由于 $f(x)$ 在区间 $[a,b]$ 上连续,根据闭区间上连续函数的最大值-最小值存在定理,$f(x)$ 在 $[a,b]$ 上必定取得它的最大值 M 和最小值 m. 这里有两种可能的情形:

情形 1 $M=m$. 则 $f(x)$ 在区间 $[a,b]$ 上必然取相同的数值 M: $f(x)=M$. 由此,对于任意 $x\in(a,b)$,都有 $f'(x)=0$. 点 c 可以是区间 (a,b) 中的任何值.

情形 2 $M>m$. 由于 $f(a)=f(b)$,则最大值 M 和最小值 m 之中至少有一个是在区间 (a,b) 内部取得. 为了确定起见,不妨设最大值 M 在区间 (a,b) 内部 c 处取得. 利用条件(2),$f(x)$ 在 c 处可导. 由费马引理,得 $f'(c)=0$.

拉格朗日中值定理 设函数 $f(x)$ 满足下面两个条件:

(1) 在闭区间 $[a,b]$ 上连续;

(2) 在开区间 (a,b) 上可导,

则在区间 (a,b) 内至少存在一点 c,使

$$\frac{f(b)-f(a)}{b-a}=f'(c),$$

或者写成

$$f(b)-f(a)=f'(c)(b-a). \tag{4.1}$$

In geometric language, the Lagrange Mean Value Theorem is easy to state and understand. It says that, if the graph of a continuous function has a non vertical tangent line at every point between A and B, then there is at least one point C on the graph between A and B at which the tangent line is parallel to the secant line AB (In Figure 4.2, there is just one such point C; in Figure 4.3, there are several).

利用几何语言,拉格朗日中值定理是很容易表述和理解的:如果一个连续函数的图像在以 A,B 为端点的区间内任意一点上都有不垂直于 x 轴的切线,那么在曲线弧上至少存在一点 C,使得该点处的切线平行于割线 AB(在图 4.2 中只存在一个这样的点,图 4.3 中就有数个这样的点).

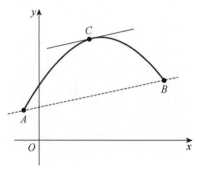

Figure 4.2
图 4.2

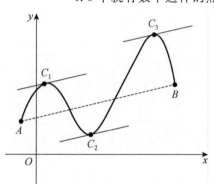

Figure 4.3
图 4.3

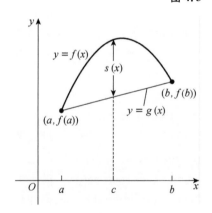

Figure 4.4
图 4.4

Proof Our proof rests on a careful analysis of the function
$$s(x) = f(x) - g(x),$$
introduced in Figure 4.4. Here $y = g(x)$ is the equation of the line through $(a, f(a))$ and $(b, f(b))$. Since this line has slope $\dfrac{f(b)-f(a)}{b-a}$ and goes through $(a, f(a))$, the point-slope form for its equation is

证明 我们的证明是建立在对图 4.4 所示函数
$$s(x) = f(x) - g(x)$$
仔细分析的基础上的. 图 4.4 中 $y=g(x)$ 是经过点 $(a,f(a))$ 和点 $(b,f(b))$ 的直线. 由于这条斜率为 $\dfrac{f(b)-f(a)}{b-a}$ 的直线经过点 $(a,f(a))$,所以它的点斜式方程为

$$g(x)-f(a)=\frac{f(b)-f(a)}{b-a}(x-a).$$

This, in turn, yields a formula for $s(x)$:
$$s(x)=f(x)-g(x)$$
$$=f(x)-f(a)-\frac{f(b)-f(a)}{b-a}(x-a).$$

Note immediately that
$$s(b)=s(a)=0$$
and that, for x in (a,b),
$$s'(x)=f'(x)-\frac{f(b)-f(a)}{b-a}.$$

Now we make a crucial observation. If we knew that there was a number c in (a,b) satisfying $s'(c)=0$, we would be all done. For then the last equation would say that
$$0=f'(c)-\frac{f(b)-f(a)}{b-a},$$
which is equivalent to the conclusion of the theorem.

To find some c in (a,b) such that $s'(c)=0$, we have the reason as follow. Clearly, $s(x)$ is continuous on $[a,b]$, being the difference of two continuous functions. Thus, by the Max-Min Existence Theorem, $s(x)$ must attain both a maximum and a minimum value on $[a,b]$, and if both the maximum value and the minimum value are 0, then $s(x)=0$, $x\in[a,b]$. Consequently, $s'(x)=0$ for all x in (a,b), much more than we need.

If either the maximum value or the minimum value is different from 0, then that value is attained at an interior point c, since $s(a)=s(b)=0$. Now $s(x)$ has a derivative at each point of (a,b), and so, by the Fermat's Lemma, $s'(c)=0$. That is all we needed to know.

At last, a more general mean value theorem is given:

The Cauchy Mean Value Theorem Let $f(x)$ and $g(x)$ be functions that fulfills following hypotheses:

(1) they are continuous on the closed interval $[a,b]$;

(2) they are differentiable on the open interval (a,b);

(3) $f'(x)$ and $g'(x)$ are not equal to zero at the same time;

(4) $g(a)\neq g(b)$.

Then, there is at least one number c in (a, b) such that
$$\frac{f'(c)}{g'(c)} = \frac{f(b) - f(a)}{g(b) - g(a)}. \qquad (4.2)$$

Remark If we take $g(x) = x$, then $g(b) - g(a) = b - a$, $g'(x) = 1$, so formula (4.2) can be written as
$$f(b) - f(a) = f'(c)(b - a).$$
it become Lagrange mean value formula (4.1).

Example 1 Find the number c satisfying the conclusion of the Mean Value Theorem for $f(x) = 2\sqrt{x}$ on $[1, 4]$.

Solution We have
$$f'(x) = 2 \times \frac{1}{2} x^{-\frac{1}{2}} = \frac{1}{\sqrt{x}},$$
and
$$\frac{f(4) - f(1)}{4 - 1} = \frac{4 - 2}{3} = \frac{2}{3}.$$
Thus, we must solve the equation
$$\frac{1}{\sqrt{c}} = \frac{2}{3}.$$
The single solution is $c = \frac{9}{4}$ (Figure 4.5).

则至少存在一点 $c \in (a, b)$,使得
$$\frac{f'(c)}{g'(c)} = \frac{f(b) - f(a)}{g(b) - g(a)}. \qquad (4.2)$$

注 如果取 $g(x) = x$,那么 $g(b) - g(a) = b - a$, $g'(x) = 1$. 于是公式(4.2)就可以写成
$$f(b) - f(a) = f'(c)(b - a).$$
它就变成了拉格朗日中值公式(4.1).

例 1 找出 $f(x) = 2\sqrt{x}$ 在区间$[1, 4]$上适合微分中值定理结论的数值 c.

解 我们有
$$f'(x) = 2 \times \frac{1}{2} x^{-\frac{1}{2}} = \frac{1}{\sqrt{x}},$$
并且有
$$\frac{f(4) - f(1)}{4 - 1} = \frac{4 - 2}{3} = \frac{2}{3}.$$
因此,我们要解方程
$$\frac{1}{\sqrt{c}} = \frac{2}{3}.$$
解得唯一解 $c = \frac{9}{4}$ (图 4.5).

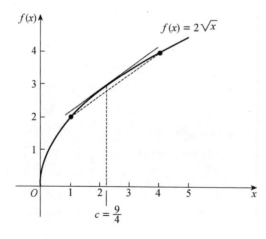

Figure 4.5
图 4.5

Example 2 Let the function $f(x) = x^3 - x^2 - x + 1$ be defined on $[-1, 2]$. Find all number c satisfying the conclusion of the Mean Value Theorem.

Solution Figure 4.6 shows a graph of the function $f(x)$. From this graph, it appears that there are two

例 2 设函数 $f(x) = x^3 - x^2 - x + 1$ 定义在$[-1, 2]$上,找出所有满足微分中值定理结论的数值 c.

解 函数 $f(x)$ 的图像如图 4.6 所示. 我们从图像中可以看出,存在两个点 c_1 和 c_2

numbers c_1 and c_2 with the required property. We now find
$$f'(x) = 3x^2 - 2x - 1,$$
and
$$\frac{f(2) - f(-1)}{2 - (-1)} = \frac{3 - 0}{3} = 1.$$
Therefore, we must solve the equation
$$3c^2 - 2c - 1 = 1,$$
or, equivalently
$$3c^2 - 2c - 2 = 0.$$
By the Quadratic Formula, we have two solutions, $\frac{2 \pm \sqrt{4 + 24}}{6}$, which correspond to $c_1 \approx -0.55$ and $c_2 \approx 1.22$. Both number are in the interval $(-1, 2)$.

符合要求. 我们求得
$$f'(x) = 3x^2 - 2x - 1$$
及
$$\frac{f(2) - f(-1)}{2 - (-1)} = \frac{3 - 0}{3} = 1.$$
因此,我们要解方程
$$3c^2 - 2c - 1 = 1,$$
即
$$3c^2 - 2c - 2 = 0.$$
运用二次根式求解法,我们求得两个解 $\frac{2 \pm \sqrt{4 + 24}}{6}$,对应的近似值是 $c_1 \approx -0.55$ 和 $c_2 \approx 1.22$. 它们都在区间 $(-1, 2)$ 内.

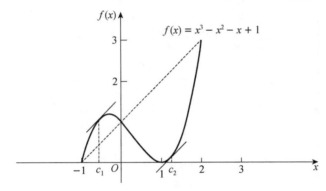

Figure 4.6
图 4.6

If the function $s(t)$ represents the position of an object at time t, then the Mean Value Theorem states that over any interval of time, there is some time for which the instantaneous velocity equals the average velocity.

Example 3 Suppose that an object has position function
$$s(t) = t^2 - t - 2.$$
Find the average velocity over the interval $[3, 6]$, and find the time at which the instantaneous velocity equals the average velocity.

Solution The average velocity over the interval $[3, 6]$ is equal to
$$\frac{s(6) - s(3)}{6 - 3} = 8.$$
The instantaneous velocity is
$$s'(t) = 2t - 1.$$

如果函数 $s(t)$ 代表物体随着时间 t 改变的路程,则微分中值定理表示在任一给定的时间间隔内,存在某时刻的瞬时速度等于这段时间间隔的平均速度.

例 3 假设一物体有路程函数
$$s(t) = t^2 - t - 2,$$
求区间 $[3, 6]$ 上的平均速度,并求在哪个时刻的瞬时速度与平均速度相等.

解 区间 $[3, 6]$ 上的平均速度等于
$$\frac{s(6) - s(3)}{6 - 3} = 8.$$

物体的瞬时速度为
$$s'(t) = 2t - 1.$$

To find the point where average velocity equals instantaneous velocity, we equate
$$8 = 2t - 1$$
and solve to get $t = \dfrac{9}{2}$.

The next theorem will be used repeatedly in this and the next chapter. In words, it says that two functions with the same derivative differ by a constant (possibly the zero constant) (Figure 4.7).

为了找到平均速度与瞬时速度相等的那个时刻,我们令
$$8 = 2t - 1,$$
并解得 $t = \dfrac{9}{2}$.

下面的定理会在后面的章节里经常用到. 它可以这样表达:两个导数相等的函数之间只相差一个常数(可能是常数 0)(图 4.7).

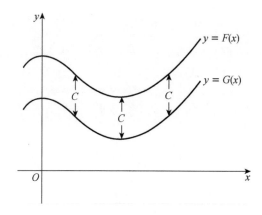

Figure 4.7
图 4.7

Theorem 4.1 If $F'(x) = G'(x)$ for all x in (a, b), then there is a constant C such that
$$F(x) = G(x) + C$$
for all x in (a, b).

Proof Let $H(x) = F(x) - G(x)$, then
$$H'(x) = F'(x) - G'(x) = 0$$
for all x in (a, b).

Choose any point x_1 in (a, b), and let x be any other point there. The function $H(x)$ satisfies the hypotheses of the Mean Value Theorem on the closed interval with end points x_1 and x. Thus, there is a number c between x_1 and x such that
$$H(x) - H(x_1) = H'(c)(x - x_1).$$
But $H'(c) = 0$ by hypothesis. Therefore,
$$H(x) - H(x_1) = 0$$
or, equivalently,
$$H(x) = H(x_1).$$

定理 4.1 如果对于 (a, b) 内的所有 x 都有 $F'(x) = G'(x)$,那么必然存在一个常数 C,使得对于 (a, b) 内的所有 x,都有
$$F(x) = G(x) + C.$$

证明 令 $H(x) = F(x) - G(x)$,则对于 (a, b) 内的所有 x,都有
$$H'(x) = F'(x) - G'(x) = 0.$$

选择 (a, b) 内任意一个点 x_1,令 x 是 (a, b) 内任意的其他点. 函数 $H(x)$ 在以 x_1, x 为两个端点的闭区间上满足微分中值定理的假定. 因此,在 x_1 与 x 之间存在一个数 c,使得
$$H(x) - H(x_1) = H'(c)(x - x_1).$$
但是根据假定,$H'(c) = 0$. 因此,对任何 $x \in (a, b)$,有
$$H(x) - H(x_1) = 0,$$
即
$$H(x) = H(x_1).$$

For all x in (a,b), since
$$H(x)=F(x)-G(x),$$
we conclude that
$$F(x)-G(x)=H(x_1).$$
Now let $C=H(x_1)$, and we have the conclusion
$$F(x)=G(x)+C.$$

4.2 The L'Hospital Rule
4.2 洛必达法则

If $x \to a$ (or $x \to \infty$), both $f(x) \to 0$ and $g(x) \to 0$ or $f(x) \to \infty$ and $g(x) \to \infty$, then $\lim\limits_{\substack{x \to a \\ (x \to \infty)}} \dfrac{f(x)}{g(x)}$ may or may not exist. We often call this limit as the **indeterminate form**, and denote them as $\dfrac{0}{0}$ or $\dfrac{\infty}{\infty}$. The limit $\lim\limits_{x \to 0} \dfrac{\sin x}{x}$ is an example of indeterminate form $\dfrac{0}{0}$. For this type of limit, we can not use the law that the limit of the quotient equals to the quotient of the limit.

In the following part, we will derive a simple and important method to evaluate this type of limit according to the Cauchy Mean Value Theorem.

We will focus on the indeterminate form $\dfrac{0}{0}$ as $x \to a$. For this case, we have the following theorem:

Theorem 4.2 If the functions $f(x)$ and $g(x)$ satisfy the following conditions:

(1) $\lim\limits_{x \to a} f(x) = 0$ and $\lim\limits_{x \to a} g(x) = 0$;

(2) $f'(x)$ and $g'(x)$ both exist and $g'(x) \neq 0$ on an open interval that contains a (except possibly at a);

(3) $\lim\limits_{x \to a} \dfrac{f'(x)}{g'(x)}$ exists (or is infinity),

then
$$\lim_{x \to a} \dfrac{f(x)}{g(x)} = \lim_{x \to a} \dfrac{f'(x)}{g'(x)}.$$

That is, when $\lim\limits_{x \to a} \dfrac{f'(x)}{g'(x)}$ exists, $\lim\limits_{x \to a} \dfrac{f(x)}{g(x)}$ also exists and equals to $\lim\limits_{x \to a} \dfrac{f'(x)}{g'(x)}$; when $\lim\limits_{x \to a} \dfrac{f'(x)}{g'(x)}$ is infinity, $\lim\limits_{x \to a} \dfrac{f(x)}{g(x)}$ is

因为对 (a,b) 内的任意 x，有
$$H(x)=F(x)-G(x),$$
所以我们得到
$$F(x)-G(x)=H(x_1).$$
令 $C=H(x_1)$，得到结论
$$F(x)=G(x)+C.$$

如果当 $x \to a$（或 $x \to \infty$）时，同时有 $f(x) \to 0$ 和 $g(x) \to 0$ 或 $f(x) \to \infty$ 和 $g(x) \to \infty$，那么极限 $\lim\limits_{\substack{x \to a \\ (x \to \infty)}} \dfrac{f(x)}{g(x)}$ 可能存在，也可能不存在。通常把这种极限叫作**未定式**，并分别简记为 $\dfrac{0}{0}$ 或 $\dfrac{\infty}{\infty}$。极限 $\lim\limits_{x \to 0} \dfrac{\sin x}{x}$ 就是 $\dfrac{0}{0}$ 未定式的一个例子。对于这类极限，不能用"商的极限等于极限的商"这一法则来求。

下面我们将根据柯西中值定理来推出求这类极限的一种简便且重要的方法。

我们着重讨论 $x \to a$ 时的 $\dfrac{0}{0}$ 未定式。关于这情形有以下定理：

定理 4.2 如果函数 $f(x), g(x)$ 满足下列条件：

(1) $\lim\limits_{x \to a} f(x) = 0$，$\lim\limits_{x \to a} g(x) = 0$;

(2) 在包含点 a 的开区间内（点 a 可能除外），$f'(x)$ 及 $g'(x)$ 都存在且 $g'(x) \neq 0$;

(3) $\lim\limits_{x \to a} \dfrac{f'(x)}{g'(x)}$ 存在（或为无穷大），

那么
$$\lim_{x \to a} \dfrac{f(x)}{g(x)} = \lim_{x \to a} \dfrac{f'(x)}{g'(x)}.$$

这就是说，当 $\lim\limits_{x \to a} \dfrac{f'(x)}{g'(x)}$ 存在时，$\lim\limits_{x \to a} \dfrac{f(x)}{g(x)}$ 也存在且等于 $\lim\limits_{x \to a} \dfrac{f'(x)}{g'(x)}$；当 $\lim\limits_{x \to a} \dfrac{f'(x)}{g'(x)}$ 为无穷

also infinity. The method which using the derivative of numerator and denominator to compute the indeterminate form is called as the **L'Hospital Rule**.

Example 1 Find $\lim\limits_{x\to 0}\dfrac{\sin ax}{\sin bx}$ $(b\neq 0)$.

Solution $\lim\limits_{x\to 0}\dfrac{\sin ax}{\sin bx}=\lim\limits_{x\to 0}\dfrac{a\cos ax}{b\cos bx}=\dfrac{a}{b}$.

Example 2 Find $\lim\limits_{x\to 1}\dfrac{x^3-3x+2}{x^3-x^2-x+1}$.

Solution $\lim\limits_{x\to 1}\dfrac{x^3-3x+2}{x^3-x^2-x+1}=\lim\limits_{x\to 1}\dfrac{3x^2-3}{3x^2-2x-1}$
$=\lim\limits_{x\to 1}\dfrac{6x}{6x-2}=\dfrac{3}{2}$.

Remark The previous expression $\lim\limits_{x\to 1}\dfrac{6x}{6x-2}$ is not indeterminate form, we cannot apply the L'Hospital Rule on it. Otherwise, it will lead to erroneous results. In the future, we should always pay attention to this point when we use the L'Hospital Rule. That is to say, if the expression is not indeterminate form, we cannot apply the L'Hospital Rule.

Example 3 Find $\lim\limits_{x\to 0}\dfrac{x-\sin x}{x^3}$.

Solution $\lim\limits_{x\to 0}\dfrac{x-\sin x}{x^3}=\lim\limits_{x\to 0}\dfrac{1-\cos x}{3x^2}$
$=\lim\limits_{x\to 0}\dfrac{\sin x}{6x}=\dfrac{1}{6}$.

For the indeterminate form $\dfrac{0}{0}$ as $x\to\infty$, and for the indeterminate form $\dfrac{\infty}{\infty}$ as $x\to a$ or $x\to\infty$, the L'Hospital Rule is also valid. For instance, for the indeterminate form $\dfrac{\infty}{\infty}$ as $x\to\infty$, we have the following theorem:

Theorem 4.3 If the functions $f(x)$ and $g(x)$ satisfy the following conditions:

(1) $\lim\limits_{x\to\infty}f(x)=\infty$ and $\lim\limits_{x\to\infty}g(x)=\infty$;

(2) $f'(x)$ and $g'(x)$ both exist and $g'(x)\neq 0$ as $|x|>N$ (N is a constant);

(3) $\lim\limits_{x\to\infty}\dfrac{f'(x)}{g'(x)}$ exist (or is infinity),

大时,$\lim\limits_{x\to a}\dfrac{f(x)}{g(x)}$ 也是无穷大. 这种在一定条件下通过分子、分母分别求导数再求极限来确定未定式的值的方法称为**洛必达法则**.

例 1 求 $\lim\limits_{x\to 0}\dfrac{\sin ax}{\sin bx}$ $(b\neq 0)$.

解 $\lim\limits_{x\to 0}\dfrac{\sin ax}{\sin bx}=\lim\limits_{x\to 0}\dfrac{a\cos ax}{b\cos bx}=\dfrac{a}{b}$.

例 2 求 $\lim\limits_{x\to 1}\dfrac{x^3-3x+2}{x^3-x^2-x+1}$.

解 $\lim\limits_{x\to 1}\dfrac{x^3-3x+2}{x^3-x^2-x+1}=\lim\limits_{x\to 1}\dfrac{3x^2-3}{3x^2-2x-1}$
$=\lim\limits_{x\to 1}\dfrac{6x}{6x-2}=\dfrac{3}{2}$.

注 上式中的 $\lim\limits_{x\to 1}\dfrac{6x}{6x-2}$ 已不是未定式,不能对它应用洛必达法则,否则会导致错误结果. 以后使用洛必达法则时,应当经常注意这一点. 也就是说,如果不是未定式,就不能应用洛必达法则.

例 3 求 $\lim\limits_{x\to 0}\dfrac{x-\sin x}{x^3}$.

解 $\lim\limits_{x\to 0}\dfrac{x-\sin x}{x^3}=\lim\limits_{x\to 0}\dfrac{1-\cos x}{3x^2}$
$=\lim\limits_{x\to 0}\dfrac{\sin x}{6x}=\dfrac{1}{6}$.

对于 $x\to\infty$ 时的 $\dfrac{0}{0}$ 未定式,以及对于 $x\to a$ 或 $x\to\infty$ 时的 $\dfrac{\infty}{\infty}$ 未定式,也有相应的洛必达法则. 例如,对于 $x\to\infty$ 时的 $\dfrac{\infty}{\infty}$ 未定式有以下定理:

定理 4.3 如果函数 $f(x)$ 和 $g(x)$ 满足下列条件:

(1) $\lim\limits_{x\to\infty}f(x)=\infty$,$\lim\limits_{x\to\infty}g(x)=\infty$;

(2) 当 $|x|>N$(N 是常数)时,$f'(x)$ 与 $g'(x)$ 都存在,且 $g'(x)\neq 0$;

(3) $\lim\limits_{x\to\infty}\dfrac{f'(x)}{g'(x)}$ 存在(或为无穷大),

then
$$\lim_{x\to\infty}\frac{f(x)}{g(x)}=\lim_{x\to\infty}\frac{f'(x)}{g'(x)}.$$

Example 4   Find $\lim\limits_{x\to+\infty}\dfrac{\dfrac{\pi}{2}-\arctan x}{\dfrac{1}{x}}$.

Solution
$$\lim_{x\to+\infty}\frac{\dfrac{\pi}{2}-\arctan x}{\dfrac{1}{x}}=\lim_{x\to+\infty}\frac{-\dfrac{1}{1+x^2}}{-\dfrac{1}{x^2}}$$
$$=\lim_{x\to+\infty}\frac{x^2}{1+x^2}=1.$$

Example 5   Find $\lim\limits_{x\to+\infty}\dfrac{\ln x}{x^n}$ $(n>0)$.

Solution
$$\lim_{x\to+\infty}\frac{\ln x}{x^n}=\lim_{x\to+\infty}\frac{\dfrac{1}{x}}{nx^{n-1}}$$
$$=\lim_{x\to+\infty}\frac{1}{nx^n}=0.$$

Example 6   Find $\lim\limits_{x\to+\infty}\dfrac{x^n}{e^{\lambda x}}$ (n is a positive integer, $\lambda>0$).

Solution   Using the L'Hospital Rule n times, we have
$$\lim_{x\to+\infty}\frac{x^n}{e^{\lambda x}}=\lim_{x\to+\infty}\frac{nx^{n-1}}{\lambda e^{\lambda x}}$$
$$=\lim_{x\to+\infty}\frac{n(n-1)x^{n-2}}{\lambda^2 e^{\lambda x}}$$
$$=\cdots=\lim_{x\to+\infty}\frac{n!}{\lambda^n e^{\lambda x}}=0.$$

There are other types of indeterminate form as $0\cdot\infty, \infty-\infty, 0^0, 1^\infty, \infty^0$. We can convert them into the indeterminate form $\dfrac{0}{0}$ or $\dfrac{\infty}{\infty}$. Here are some examples:

Example 7   Find $\lim\limits_{x\to 0^+} x^n \ln x$ $(n>0)$.

Solution   This is an indeterminate form $0\cdot\infty$. We have
$$x^n\ln x=\frac{\ln x}{\dfrac{1}{x^n}}.$$

The right hand side of above equation is an indeterminate form $\dfrac{\infty}{\infty}$ as $x\to 0^+$. Applying the L'Hospital Rule, we get

$$\lim_{x\to 0^+} x^n \ln x = \lim_{x\to 0^+} \frac{\ln x}{x^{-n}} = \lim_{x\to 0^+} \frac{\frac{1}{x}}{-nx^{-n-1}}$$
$$= \lim_{x\to 0^+} \left(-\frac{x^n}{n}\right) = 0.$$

Example 8 Find $\lim\limits_{x\to \pi/2}(\sec x - \tan x)$.

Solution This is an indeterminate form $\infty - \infty$. We have
$$\sec x - \tan x = \frac{1-\sin x}{\cos x}.$$

The right hand side of above equation is an indeterminate form $\frac{0}{0}$ as $x\to \frac{\pi}{2}$. Applying the L'Hospital Rule, we get
$$\lim_{x\to \pi/2}(\sec x - \tan x) = \lim_{x\to \pi/2}\frac{1-\sin x}{\cos x}$$
$$= \lim_{x\to \pi/2}\frac{-\cos x}{-\sin x} = 0.$$

Example 9 Find $\lim\limits_{x\to 0^+} x^x$.

Solution This is an indeterminate form 0^0. Let $y = x^x$, take logarithm, then
$$\ln y = x \ln x.$$
The right hand side of above equation is indeterminate form $0 \cdot \infty$ as $x \to 0^+$. Apply to the result of Example 7, we have
$$\lim_{x\to 0^+} \ln y = \lim_{x\to 0^+} x \ln x = 0.$$
Since $y = e^{\ln y}$, thus
$$\lim_{x\to 0^+} y = \lim_{x\to 0^+} e^{\ln y} = e^{\lim\limits_{x\to 0^+} \ln y},$$
so
$$\lim_{x\to 0^+} x^x = \lim_{x\to 0^+} y = e^0 = 1.$$

At last we remark that the L'Hospital Rule give us a method to compute indeterminate form. When the conditions in the theorem are satisfied, the limit must exist (or infinity), but when the conditions in the theorem are not satisfied, the limit may exist, that is when $\lim \frac{f'(x)}{g'(x)}$ does not exist (except infinity), $\lim \frac{f(x)}{g(x)}$ may exist.

4.3 The Criterion of the Monotonicity of Functions
4.3 函数的单调性判别法

1. The First Derivative and Monotonicity

The first derivative $f'(x)$ give us the slope of the tangent line to the graph of the function $f(x)$ at the point x. Thus, if $f'(x)>0$, then the tangent line is rising to the right, suggesting that $f(x)$ is increasing. Similarly, if $f'(x)<0$, then the tangent line is falling to the right, suggesting that $f(x)$ is decreasing (Figure 4.8).

1. 函数的一阶导数与单调性

一阶导数 $f'(x)$ 为我们提供了函数 $f(x)$ 的图像在点 x 处的切线斜率. 因此, 如果 $f'(x)>0$, 则切线就向右边上升. 这意味着 $f(x)$ 单调递增. 类似地, 如果 $f'(x)<0$, 则切线就向右边下降. 这意味着 $f(x)$ 单调递减 (图 4.8).

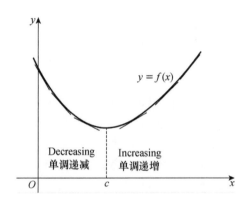

Figure 4.8
图 4.8

Theorem 4.4 (The Monotonicity Theorem) Let $f(x)$ be continuous on an interval I and differentiable at every interior point of I.

(1) If $f'(x)>0$ for all x interior to I, then $f(x)$ is increasing on I.

(2) If $f'(x)<0$ for all x interior to I, then $f(x)$ is decreasing on I.

Proof Consider any two points x_1 and x_2 of I with $x_1<x_2$. By the Mean Value Theorem applied to the interval $[x_1, x_2]$, there is a number c in (x_1, x_2) satisfying
$$f(x_2) - f(x_1) = f'(c)(x_2 - x_1).$$
Since $f'(c)>0$, we see that
$$f(x_2) - f(x_1) > 0,$$
that is

定理 4.4 (单调性定理) 设 $f(x)$ 在区间 I 上连续并且在 I 内每一点都可导.

(1) 如果对于 I 内的所有 x, 都有 $f'(x)>0$, 那么 $f(x)$ 在 I 上单调递增;

(2) 如果对于 I 内的所有 x, 都有 $f'(x)<0$, 那么 $f(x)$ 在 I 上单调递减.

证明 考虑 I 上的任意两点 x_1, x_2, 它们的关系是 $x_1<x_2$. 在区间 $[x_1, x_2]$ 上运用微分中值定理, 则在 (x_1, x_2) 内存在数 c, 使得
$$f(x_2) - f(x_1) = f'(c)(x_2 - x_1).$$
由于 $f'(c)>0$, 得到
$$f(x_2) - f(x_1) > 0,$$
也就是说

$$f(x_2) > f(x_1).$$

So $f(x)$ is increasing on I.

The case where $f'(x) < 0$ on I is handled similarly.

This theorem usually allows us to determine precisely where a differentiable function increase and where it decrease. It is a matter of solving two inequalities.

Example 1 Let
$$f(x) = 2x^3 - 3x^2 - 12x + 7.$$
Find where $f(x)$ is increasing and where it is decreasing.

Solution We begin by finding the derivative of $f(x)$:
$$f'(x) = 6x^2 - 6x - 12 = 6(x+1)(x-2).$$
We need to determine where
$$(x+1)(x-2) > 0,$$
and also where
$$(x+1)(x-2) < 0.$$

The split points are -1 and 2. They split the x-axis into three intervals: $(-\infty, -1)$, $(-1, 2)$ and $(2, +\infty)$. Using the test points $-2, 0$ and 3, we conclude that $f'(x) > 0$ on the first and last of these intervals and that $f'(x) < 0$ on the middle interval (Figure 4.9). Thus, by Theorem 4.4, $f(x)$ is increasing on $(-\infty, -1]$ and $[2, +\infty)$; it is decreasing on $[-1, 2]$. Note that these intervals include the end points of these intervals, even though $f'(x) = 0$ at those points. This is because that $f(x)$ is continuous at those points. The graph of the function $f(x)$ is shown in Figure 4.10.

所以 $f(x)$ 在区间 I 上单调递增.

对于在 I 上 $f'(x) < 0$ 的情况，可以用类似的方法证明.

这个定理能让我们准确地判断一个可导函数的递增区间和递减区间. 它是一个解两个不等式的问题.

例 1 设
$$f(x) = 2x^3 - 3x^2 - 12x + 7,$$
求出 $f(x)$ 的递增区间和递减区间.

解 我们首先求出 $f(x)$ 的导数：
$$f'(x) = 6x^2 - 6x - 12 = 6(x+1)(x-2).$$
我们需要判定在何处有
$$(x+1)(x-2) > 0,$$
在何处有
$$(x+1)(x-2) < 0.$$

分隔点是 -1 和 2. 它们将 x 轴分隔成三个区间：$(-\infty, -1)$，$(-1, 2)$ 和 $(2, +\infty)$. 用测试点 $-2, 0, 3$，推断出在第一个和最后一个区间内 $f'(x) > 0$，在中间的区间内 $f'(x) < 0$（图 4.9）. 因此，根据定理 4.4，$f(x)$ 在 $(-\infty, -1]$ 和 $[2, +\infty)$ 上单调递增，在 $[-1, 2]$ 上单调递减. 注意到这些区间包括端点，尽管在那些端点处 $f'(x) = 0$. 这是因为 $f(x)$ 在那些端点处连续. 函数 $f(x)$ 的图像如图 4.10 所示.

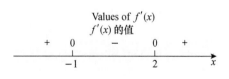

Figure 4.9
图 4.9

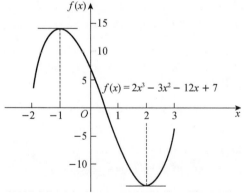

Figure 4.10
图 4.10

Example 2 Determine where $f(x)=x/(1+x^2)$ is increasing and where it is decreasing.

Solution We have
$$f'(x)=\frac{(1+x^2)-x\cdot 2x}{(1+x^2)^2}=\frac{1-x^2}{(1+x^2)^2}$$
$$=\frac{(1+x)(1-x)}{(1+x^2)^2}.$$

Since the denominator is always positive, $f'(x)$ has the same sign as the numerator $(1+x)(1-x)$. The split points, -1 and 1, determine the three intervals: $(-\infty,-1)$, $(-1,1)$ and $(1,+\infty)$. When we test them, we find that $f'(x)<0$ on the first and last of these intervals and that $f'(x)>0$ on the middle interval (Figure 4.11). From Theorem 4.4, We conclude that $f(x)$ is decreasing on $(-\infty,-1]$ and $[1,+\infty)$ and that it is increasing on $[-1,1]$. The graph of the function $f(x)$ is shown in Figure 4.12.

例 2 求出 $f(x)=x/(1+x^2)$ 的递增区间和递减区间.

解 我们有
$$f'(x)=\frac{(1+x^2)-x\cdot 2x}{(1+x^2)^2}=\frac{1-x^2}{(1+x^2)^2}$$
$$=\frac{(1+x)(1-x)}{(1+x^2)^2}.$$

因为分母恒为正数,所以 $f'(x)$ 的符号与分子 $(1+x)(1-x)$ 相同.分隔点 -1 和 1 决定了三个区间 $(-\infty,-1)$,$(-1,1)$ 和 $(1,+\infty)$.我们测试时发现,在第一个和最后一个区间内 $f'(x)<0$,在中间的区间内 $f'(x)>0$(图 4.11).根据定理 4.4,我们推断出 $f(x)$ 在 $(-\infty,-1]$ 和 $[1,+\infty)$ 上单调递减,而在 $[-1,1]$ 上单调递增.函数 $f(x)$ 的图像如图 4.12 所示.

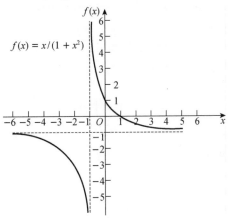

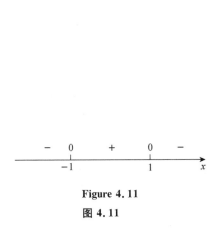

Figure 4.11
图 4.11

Figure 4.12
图 4.12

2. The Second Derivative and Concavity

A function may be increasing and still have a very wiggly graph (Figure 4.13). To analyze wiggles, we need to study how the tangent line turns as we move from left to right along the graph. If the tangent line turns steadily in the counter clockwise direction, we say that the graph is **concave up**; If the tangent turns in the clockwise direction, the graph is **concave down**. Both definitions are better stated in terms of functions and their derivatives.

2. 二阶导数和凹性

一个函数可能是单调递增的,但依然会具有摆动的图像(图 4.13).为了分析图像的摆动,我们需要研究沿着图像从左到右移动时它的切线怎样转动.如果切线稳定地沿逆时针方向转动,我们就说图像**上凹**;如果切线稳定地沿顺时针方向转动,我们就说图像**下凹**.这两个定义都可用函数及其导数给出较好的表达.

Chapter 4　Applications of the Derivative
第 4 章　导数的应用

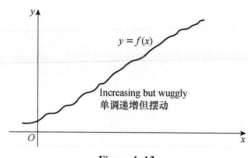

Figure 4.13

图 4.13

Definition 4.1　Let the function $f(x)$ be differentiable on an open interval I. We say that $f(x)$ (as well as its graph) is **concave up** on I if $f'(x)$ is increasing on I (Figure 4.14(a)), and we say that is **concave down** on I if $f'(x)$ is decreasing on I (Figure 4.14(b)).

Definition 4.2　Let function $f(x)$ be continuous on the interval (a,b), and (a,b) contains the point x_0. x_0 is called an **inflection point** if $f(x)$ is concave up (or concave down) on (a,x_0), and concave down (or concave up) on (x_0,b) (Figure 4.14(c)).

定义 4.1　设函数 $f(x)$ 在开区间 I 内可导. 如果 $f'(x)$ 在 I 内单调递增, 则我们称 $f(x)$（和它的图形）在 I 内**上凹**（图 4.14(a)）；如果 $f'(x)$ 在 I 内单调递减, 则我们称 $f(x)$ 在 I 内**下凹**（图 4.14(b)）.

定义 4.2　设函数 $f(x)$ 在包含 x_0 的区间 (a,b) 内连续. 如果 $f'(x)$ 在 (a,x_0) 内上凹（或下凹）, 在 (x_0,b) 内下凹（或上凹）, 则称 x_0 为曲线 $y=f(x)$ 的**拐点**（图 4.14(c)）.

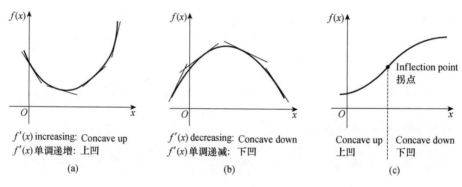

Figure 4.14

图 4.14

In view of Theorem 4.4, we have a simple criterion for deciding where a curve is concave up and where it is concave down. We simply keep in mind that the second derivative of $f(x)$ is the first derivative of $f'(x)$. Thus, $f'(x)$ is increasing if $f''(x)$ is positive; it is decreasing if $f''(x)$ is negative.

Theorem 4.5 (The Concavity Theorem)　Let the function $f(x)$ be twice differentiable on an open interval I.

基于定理 4.4, 我们有一个简单的标准来判断一条曲线的上凹区间和下凹区间. 我们很容易记住, $f(x)$ 的二阶导数是 $f'(x)$ 的一阶导数. 这样, 如果 $f''(x)$ 为正的, 则 $f'(x)$ 单调递增; 如果 $f''(x)$ 为负的, 则 $f'(x)$ 单调递减.

定理 4.5（凹性判定定理）　设函数 $f(x)$ 在开区间 I 内存在二阶导数.

(1) If $f''(x) > 0$ for all x in I, then $f(x)$ is concave up on I.

(2) If $f''(x) < 0$ for all x in I, then $f(x)$ is concave down on I.

For most functions, this theorem reduces the problem of determining concavity to the problem of solving inequalities.

Example 3 Where is the function
$$f(x) = \frac{1}{3}x^3 - x^2 - 3x + 4$$
increasing, decreasing, concave up, and concave down?

Solution We have
$$f'(x) = x^2 - 2x - 3 = (x+1)(x-3),$$
$$f''(x) = 2x - 2 = 2(x-1).$$
By solving the inequalities
$$(x+1)(x-3) > 0 \quad \text{and} \quad (x+1)(x-3) < 0,$$
we conclude that $f(x)$ is increasing on $(-\infty, -1]$ and $[3, +\infty)$ and decreasing on $[-1, 3]$ (Figure 4.15). Similarly, solving
$$2(x-1) > 0 \quad \text{and} \quad 2(x-1) < 0$$
shows that $f(x)$ is concave up on $(1, +\infty)$ and concave down on $(-\infty, 1)$. The graph of function $f(x)$ is shown in Figure 4.16.

（1）如果对于 I 内的所有 x，有 $f''(x) > 0$，那么 $f(x)$ 在 I 内上凹；

（2）如果对于 I 内的所有 x，有 $f''(x) < 0$，那么 $f(x)$ 在 I 内下凹.

对于大多数函数，这个定理把判定凹性问题简化成解不等式的问题.

例 3 函数
$$f(x) = \frac{1}{3}x^3 - x^2 - 3x + 4$$
在何处单调递增、单调递减、上凹、下凹？

解 我们有
$$f'(x) = x^2 - 2x - 3 = (x+1)(x-3),$$
$$f''(x) = 2x - 2 = 2(x-1).$$
通过解不等式
$$(x+1)(x-3) > 0 \quad \text{和} \quad (x+1)(x-3) < 0,$$
推断出 $f(x)$ 在 $(-\infty, -1]$ 和 $[3, +\infty)$ 上单调递增，在 $[-1, 3]$ 上单调递减（图 4.15）. 类似地，解
$$2(x-1) > 0 \quad \text{和} \quad 2(x-1) < 0,$$
得出 $f(x)$ 在 $(1, +\infty)$ 内上凹，在 $(-\infty, 1)$ 内下凹. 函数 $f(x)$ 的图像如图 4.16 所示.

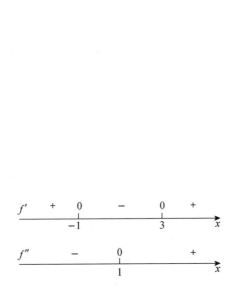

Figure 4.15
图 4.15

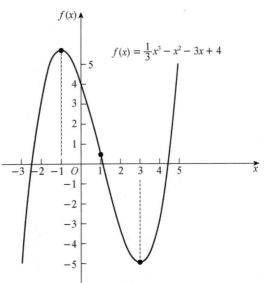

Figure 4.16
图 4.16

4.4 Maxima and Minima
4.4 最大值和最小值

Often in life, we are faced with the problem of finding the best way to do something. For example, a farmer wants to choose the mix of crops that is likely to produce the largest profit; a manufacturer would like to minimize the cost of distributing its products. Often such a problem can be formulated so that it involves maximizing or minimizing a function over a specified set. If so, the methods of calculus provide a powerful tool for solving the questions.

Definition 4.3 The function we want to maximize or minimize is called the **objective function**.

1. The Existence Question

Does the function $f(x)$ have a maximum (or minimum) value on the interval I?

There is a nice theorem that answers the max-min existence question for many of the problem that come up in practice. It is the Max-Min Existence Theorem introduced in Chapter 2:

Theorem 4.6 (The Max-Min Existence Theorem)

If function $f(x)$ is continuous on a closed interval $[a,b]$, then $f(x)$ attains both a maximum value and a minimum value there.

Note the key words in the theorem: $f(x)$ is required to be "continuous" and the interval is required to be a "closed interval".

2. Where Do Extreme Values Occur?

Usually, the objective function $f(x)$ will have an interval I as its domain. But this interval may be any of the nine types discussed in Inequalities and Absolute Values. Some of them contain their end points, some do not. For instance, $I=[a,b]$ contains both its end points; $[a,b)$ contains only its left end point; (a,b) contains neither end point. Extreme values of functions defined on closed intervals often occur at end points (Figure 4.17).

在日常生活中,我们常常会遇到寻找解决某件事件的最佳方法的问题. 例如,农民想要选择农作物的混合以获得最大的利润;生产商会想让产品的成本最低. 这类问题经常会被描述成在一个指定集合里求一个函数的最大值或最小值的问题. 若这样,微积分方法就为解决这类问题提供了一种强大的工具.

定义 4.3 我们把要求其最大值或最小值的函数叫作**目标函数**.

1. 存在性问题

函数 $f(x)$ 在区间 I 上是否存在最大值(或者最小值)?

这里有一个很好的定理,它能回答很多来自实际的关于最大值-最小值存在性的问题. 它就是在第二章介绍的最大值-最小值存在定理:

定理 4.6(最大值-最小值存在定理)

如果函数 $f(x)$ 在闭区间 $[a,b]$ 上连续,那么 $f(x)$ 在该区间上存在最大值和最小值.

注意该定理中的关键词: $f(x)$ 要求是"连续"的,区间要求是"闭区间".

2. 最值在哪里出现?

通常目标函数 $f(x)$ 会有一个区间 I 作为定义域. 但这个区间可以是"不等式与绝对值"内容中讨论的9种区间中的任何一种,有的包含它们的端点,有的不包含它们的端点. 例如,$I=[a,b]$ 包含两个端点,$[a,b)$ 只包含一个左端点,(a,b) 没有包含端点. 定义在闭区间的函数的最值经常出现在端点处(图4.17).

4.4　Maxima and Minima
4.4　最大值和最小值

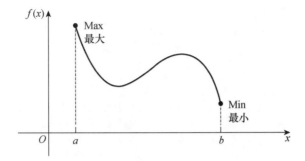

Figure 4.17
图 4.17

If c is a point at which $f'(c)=0$, we call c a **stationary point**. The name derives from the fact that the graph of $f(x)$ levels off at a stationary point, since the tangent line is horizontal. Extreme values often occur at stationary points (Figure 4.18).

如果 c 是一个使 $f'(c)=0$ 的点,则我们称 c 为**驻点**.这个名字来源于 $f(x)$ 的图像在驻点上变得水平这个事实,因为该点处的切线是水平的.最值通常也会出现在驻点处(图 4.18).

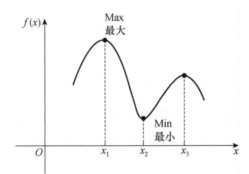

Figure 4.18
图 4.18

Finally, if c is an interior point of interval I where $f'(x)$ fails to exist, we call c a **singular point**. It is a point where the graph of $f(x)$ has a sharp corner, a vertical tangent, or perhaps takes a jump, or near where the graph wiggles very badly. Extreme values can occur at singular points (Figure 4.19), though in practical problems this is quite rare.

These three kinds of points (end points, stationary points, and singular points) are the key points of max-min theory. Any point of one of these three types in the domain of a function is called a critical point of $f(x)$.

最后,如果 c 是区间 I 内的一个使 $f'(x)$ 不存在的点,那么我们称 c 是**奇点**.这个点是使 $f(x)$ 的图像出现尖角、垂直的切线、跳跃,或者图像在该点的附近摆动很厉害的点.最值也可以出现在奇点处(图 4.19),尽管这种情况在实际问题中很少见.

这三类点(端点、驻点和奇点)是最大值-最小值理论的关键点.在函数 $f(x)$ 的定义域中,这三类点中的任意一点都称为 $f(x)$ 的临界点.

Chapter 4 Applications of the Derivative
第 4 章 导数的应用

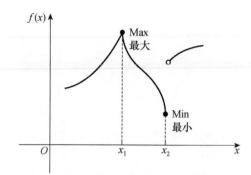

Figure 4.19
图 4.19

Example 1 Find the critical points of the function $f(x) = -2x^3 + 3x^2$ on $\left[-\dfrac{1}{2}, 2\right]$.

Solution The end points are $-\dfrac{1}{2}$ and 2. To find the stationary points, we solve the equation
$$f'(x) = -6x^2 + 6x = 0$$
for x, obtaining 0 and 1. There are no singular points. Thus, the critical points are $-\dfrac{1}{2}$, 0, 1, and 2.

Theorem 4.7 (The Critical Point Theorem) Let the function $f(x)$ be defined on an interval I containing the point c. If $f(c)$ is an extreme value, then c must be a critical point, that is, either c is

(1) an end point of I;

(2) a stationary point of $f(x)$, that is a point where $f'(c) = 0$; or

(3) a singular point of $f(x)$, that is a point where $f'(c)$ does not exist.

Proof Consider first the case where $f(c)$ is the maximum value of $f(x)$ on I, and suppose that c is neither an end point nor a singular point. We must show that c is a stationary point.

Now, since $f(c)$ is the maximum value, $f(x) \leqslant f(c)$ for all x in I, that is
$$f(x) - f(c) \leqslant 0.$$
Thus, if $x < c$, then

例 1 求函数 $f(x) = -2x^3 + 3x^2$ 在 $\left[-\dfrac{1}{2}, 2\right]$ 上的临界点.

解 端点是 $-\dfrac{1}{2}$ 和 2. 为了求驻点，我们解关于 x 的方程
$$f'(x) = -6x^2 + 6x = 0,$$
得到 0 和 1. 没有奇点. 因此，临界点是 $-\dfrac{1}{2}$，0,1 和 2.

定理 4.7 (临界点定理) 设函数 $f(x)$ 定义在包含点 c 的区间 I 上. 如果 $f(c)$ 是一个最值，那么 c 一定是临界点. 也就是说，c 是下列三种情况之一：

(1) I 的一个端点；

(2) $f(x)$ 的一个驻点，即一个使 $f'(c) = 0$ 的点；

(3) $f(x)$ 的一个奇点，即一个使 $f'(c)$ 不存在的点.

证明 首先考虑 $f(c)$ 是 $f(x)$ 在 I 上的最大值的情况. 假设 c 既不是端点又不是奇点. 我们必须证明 c 是驻点.

现在，因为 $f(c)$ 是最大值，所以对于 I 上的所有 x，有 $f(x) \leqslant f(c)$，即
$$f(x) - f(c) \leqslant 0.$$
这样，如果 $x < c$，那么

$$\frac{f(x)-f(c)}{x-c} \geqslant 0; \qquad (4.3)$$

whereas if $x>c$, then

$$\frac{f(x)-f(c)}{x-c} \leqslant 0. \qquad (4.4)$$

But $f'(c)$ exists because c is not a singular point. Consequently, when we let $x \to c^-$ in (4.3) and $x \to c^+$ in (4.4), we obtain, respectively, $f'(c) \geqslant 0$ and $f'(c) \leqslant 0$. We concluded that $f'(c)=0$, as desired.

The case where $f(c)$ is the minimum value is handled similarly.

In the proof just given, we used the fact that the inequality is preserved under the operation of taking limits.

3. How to Find Extreme Values?

In view of the Max-Min Existence Theorem and the Critical Point Theorem, we can now state a very simple procedure for finding the maximum and minimum values of a continuous function $f(x)$ on a closed interval I.

Step 1 Find the critical points of $f(x)$ on I.

Step 2 Evaluate $f(x)$ at each of these critical points. The largest of these values is the maximum value; the smallest is the minimum value.

Example 2 Find the maximum and minimum values of $f(x)=x^3$ on $[-2,2]$.

Solution The derivative of $f(x)$ is
$$f'(x)=3x^2,$$
which is defined on $(-2,2)$ and is zero only when $x=0$. The critical points therefore are $x=0$ and the end points $x=-2$ and $x=2$. Evaluating $f(x)$ at the critical points yields
$$f(-2)=-8, \quad f(0)=0, \quad \text{and} \quad f(2)=8.$$
Thus, the maximum value of $f(x)$ is 8 (attained at $x=2$) and the minimum value is -8 (attained at $x=-2$).

Remark In Example 2, $f'(0)=0$, but $f(x)$ did not attain a minimum or maximum value at $x=0$. This does not contradict the Critical Point Theorem. This theorem does not say that if c is a critical point, then $f(c)$ is a minimum or maximum value; it says that if $f(c)$ is a minimum or maximum value, then c is a critical point.

Example 3 Find the maximum and minimum values of the function $f(x) = -2x^3 + 3x^2$ on $\left[-\dfrac{1}{2}, 2\right]$.

Solution In Example 1, we identified $-\dfrac{1}{2}$, 0, 1 and 2 as the critical points. Now
$$f\left(-\dfrac{1}{2}\right) = 1,\ f(0) = 0,\ f(1) = 1,\quad \text{and}\quad f(2) = -4.$$
Thus, the maximum value is 1 (attained at both $x = -\dfrac{1}{2}$ and $x = 1$), and the minimum value is -4 (attained at $x = 2$). The graph of the function $f(x)$ is shown in Figure 4.20.

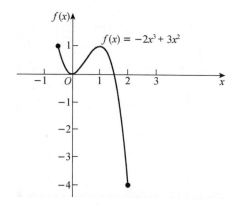

Figure 4.20

Example 4 The function $f(x) = x^{\frac{2}{3}}$ is continuous on $(-\infty, +\infty)$. Find its maximum and minimum values on $[-1, 2]$.

Solution The derivative of $f(x)$ is
$$f'(x) = \dfrac{2}{3} x^{-\frac{1}{3}},$$
which is never 0. However, $f'(0)$ does not exist, so 0 is a critical point, as the end points -1 and 2. Now
$$f(-1) = 1,\quad f(0) = 0,\quad \text{and}\quad f(2) = \sqrt[3]{4} \approx 1.59.$$
Thus, the maximum value is $\sqrt[3]{4}$; the minimum value is 0. The graph of the function $f(x)$ is shown in Figure 4.21.

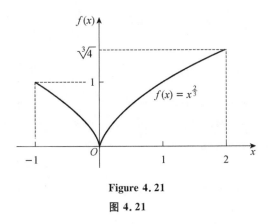

Figure 4.21
图 4.21

4.5 Local Extrema and Local Extrema on Open Intervals
4.5 局部极值与开区间上的局部极值

We recall from before section that the maximum value (if it exists) of a function on a set I is the largest value that attains on the whole set I. It is sometimes referred to as the **global maximum value**, or the **absolute maximum value** of $f(x)$. Thus, for the function $f(x)$ with domain $I=[a,b]$ whose graph is sketched in Figure 4.22, $f(a)$ is the global maximum value. But what is about $f(c)$? It is not the global maximum value, but it is the largest value that attains in a small area. We call it a **local maximum value**, or a **relative maximum value**. Of course, a global maximum value is automatically a local maximum value. Figure 4.23 illustrates a number of possibilities. Note that the global maximum value (if it exists) is simply the largest of the local maximum values. Similarly, the global minimum value is the smallest of the local minimum values.

我们回顾一下上一节的内容：函数 $f(x)$ 在区间 I 的最大值（如果存在的话）是 $f(x)$ 在 I 上所能取得的最大值. 它有时被当作**全局最大值**或**绝对最大值**来讨论. 因此，对于以 $I=[a,b]$ 为定义域、图像如图 4.22 所示的函数 $f(x)$, $f(a)$ 是全局最大值. 但 $f(c)$ 又是什么呢？它不是全局最大值，但它是某一小范围内的最大值. 我们把它称为**局部极大值**，**或相对最大值**. 当然，一个全局最大值自然也是局部极大值. 图 4.23 示意了一些可能情况. 注意全局最大值（如果存在的话）一定是局部极大值中最大的. 同样，全局最小值是局部极小值中最小的.

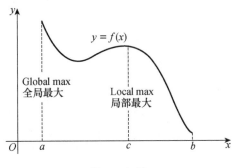

Figure 4.22
图 4.22

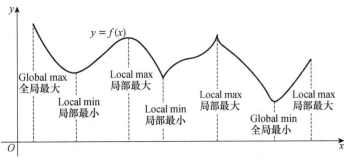

Figure 4.23
图 4.23

Here is the formal definition of local maxima and local minima:

Definition 4.4 Let D, the domain of $f(x)$, contain the point c.

(1) $f(c)$ is called a **local maximum value** of $f(x)$ if there is an interval (a,b) containing c such that $f(c)$ is the maximum value of $f(x)$ on $(a,b) \bigcap D$, and c is called a **local maximum value point** of $f(x)$.

(2) $f(c)$ is called a **local minimum value** of $f(x)$ if there is an interval (a,b) containing c such that $f(c)$ is the minimum value of $f(x)$ on $(a,b) \bigcap D$, and c is called a **local minimum value point** of $f(x)$.

The local maximum value and the local minimum value are collectively called the **local extreme value**.

1. Where Do Local Extreme Values Occur?

The Critical Point Theorem holds with the phrase extreme value replaced by local extreme value; the proof is essentially the same. Thus, the critical points (end points, stationary points, and singular points) are the candidates for points where local extrema may occur. We say candidates because we are not claiming that there must be a local extremum at every critical point. The left graph in Figure 4.24 makes this clear. However, if the derivative is positive on one side of the critical point and negative on the other (and if the function is continuous), then we have a local extremum, as shown in the middle and right graphs of Figure 4.24.

下面是局部极大值和局部极小值的正规定义：

定义4.4 设点 c 是函数 $f(x)$ 的定义域 D 内的一点.

(1) 如果在区间 (a,b) 内存在一点 c, 使得 $f(c)$ 为 $f(x)$ 在 $(a,b) \bigcap D$ 上的最大值, 则称 $f(c)$ 为 $f(x)$ 的**局部极大值**, 并称 c 为 $f(x)$ 的**局部极大值点**；

(2) 如果在区间 (a,b) 内存在一点 c, 使得 $f(c)$ 为 $f(x)$ 在 $(a,b) \bigcap D$ 上的最小值, 则称 $f(c)$ 为 $f(x)$ 的**局部极小值**, 并称 c 为 $f(x)$ 的**局部极小值点**.

局部极大值和局部极小值统称为**局部极值**.

1. 局部极值存在于何处？

将临界点定理中的"最值"这个词换成"局部极值", 定理结论同样成立, 其证明本质上是一样的. 因此, 临界点（端点、驻点和奇点）是可能取到局部极值的候选点. 说是候选点, 是因为我们并不要求在每一个临界点处都必须取到局部极值. 图4.24中左边的图形使这一点非常清晰. 但是, 如果导数在临界点的一侧为正的, 另一侧为负的（如果函数是连续的）, 就得到一个局部极值点, 如图4.24中间和右边的图形所示.

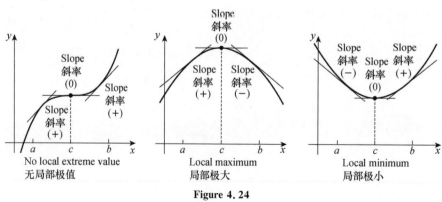

Figure 4.24

图 4.24

Theorem 4.8 (The First Derivative Test) Let $f(x)$ be continuous on an open interval (a,b) that contains a critical point c.

(1) If $f'(x)>0$ for all x in (a,c) and $f'(x)<0$ for all x in (c,b), then $f(c)$ is a local maximum value of $f(x)$.

(2) If $f'(x)<0$ for all x in (a,c) and $f'(x)>0$ for all x in (c,b), then $f(c)$ is a local minimum value of $f(x)$.

(3) If $f'(x)$ has the same sign on both sides of c, then $f(c)$ is not a local extreme value of $f(x)$.

Proof (1) Since $f'(x)>0$ for all x in (a,c), $f(x)$ is increasing on $(a,c]$ by the Monotonicity Theorem. Again, since $f'(x)<0$ for all x in (c,b), $f(x)$ is decreasing on $[c,b)$. Thus, $f(x)<f(c)$ for all x in (a,b), except of course at $x=c$. We conclude that $f(c)$ is a local maximum.

The proofs of (2) and (3) are similar.

Example 1 Find the local extreme values of the function $f(x)=x^2-6x+5$ on $(-\infty,+\infty)$.

Solution The polynomial $f(x)$ is continuous everywhere, and its derivative,
$$f'(x)=2x-6,$$
exists for all x. Thus, the only critical point for $f(x)$ is the single solution of $f'(x)=0$, that is $x=3$.

Since
$$f'(x)=2(x-3)<0$$
for $x<3$, $f(x)$ is decreasing on $(-\infty,3]$; And since
$$f'(x)=2(x-3)>0$$
for $x>3$, $f(x)$ is increasing on $[3,+\infty]$. Therefore, by the First Derivative Test, $f(3)=-4$ is a local minimum value of $f(x)$. Since $x=3$ is the only critical point, there are no other extreme values. The graph of the function $f(x)$ is shown in Figure 4.25. Note that $f(3)$ is actually the (global) minimum value in this case.

定理 4.8（一阶导数判别法） 设 $f(x)$ 在含有临界点 c 的开区间 (a,b) 内连续.

（1）如果对于 (a,c) 内的所有点 x 都有 $f'(x)>0$，且对于 (c,b) 内的所有点 x 都有 $f'(x)<0$，那么 $f(c)$ 为 $f(x)$ 的一个局部极大值；

（2）如果对于 (a,c) 内的所有点 x 都有 $f'(x)<0$，且对于 (c,b) 内的所有点 x 都有 $f'(x)>0$，那么 $f(c)$ 为 $f(x)$ 的一个局部极小值；

（3）如果 $f'(x)$ 在 c 的两侧符号相同，那么 $f(c)$ 不是 $f(x)$ 的局部极值.

证明 （1）因为对于 (a,c) 内的所有点都有 $f'(x)>0$，所以根据单调性定理，$f(x)$ 在 $(a,c]$ 上单调递增. 又因为对于 (c,b) 内的所有点都有 $f'(x)<0$，所以 $f(x)$ 在 $[c,b)$ 上单调递减. 因此，对于 (a,b) 内除 $x=c$ 的所有点都有 $f(x)<f(c)$，从而得出 $f(c)$ 就是局部极大值.

类似地，可证（2）和（3）.

例 1 求函数 $f(x)=x^2-6x+5$ 在 $(-\infty,+\infty)$ 上的局部极值.

解 多项式 $f(x)$ 是处处连续的，且它的导数
$$f'(x)=2x-6$$
对于所有点 x 都存在. 因此，唯一的临界点就是 $f'(x)=0$ 的唯一解，即 $x=3$.

因为对于 $x<3$，
$$f'(x)=2(x-3)<0,$$
所以 $f(x)$ 在 $(-\infty,3]$ 上单调递减；因为对于 $x>3$，
$$f'(x)=2(x-3)>0,$$
所以 $f(x)$ 在 $[3,+\infty)$ 上单调递增. 因此，根据一阶导数判别法，$f(3)=-4$ 为 $f(x)$ 的局部极小值. 因为 $x=3$ 是唯一的临界点，所以没有其他的局部极值. 函数 $f(x)$ 的图像如图 4.25 所示. 注意，在这种情况下，$f(3)$ 其实也是（全局）最小值.

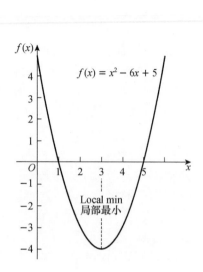

Figure 4.25
图 4.25

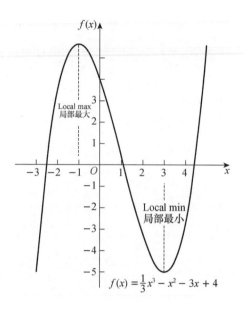

Figure 4.26
图 4.26

Example 2 Find the local extreme values of the function $f(x)=\frac{1}{3}x^3-x^2-3x+4$ on $(-\infty,+\infty)$.

Solution Since
$$f'(x)=x^2-2x-3=(x+1)(x-3),$$
the only critical points of $f(x)$ are -1 and 3. When we use the test points $-2, 0$, and 4, we learn that
$$f'(x)=(x+1)(x-3)>0$$
on $(-\infty,-1)$ and $(3,+\infty)$ and
$$f'(x)=(x+1)(x-3)<0$$
on $(-1,3)$. By the First Derivative Test, we conclude that $f(-1)=\frac{17}{3}$ is a local maximum value and that $f(3)=-5$ is a local minimum value (Figure 4.26).

There is another test for local maxima and minima that is sometimes easier to apply than the First Derivative Test. It involves evaluating the second derivative at the stationary points. It does not apply to singular points.

Theorem 4.9 (The Second Derivative Test) Let $f'(x)$ and $f''(x)$ exist at every point in an open interval (a,b) containing c, and suppose that $f'(c)=0$.

(1) If $f''(c)<0$, then $f(c)$ is a local maximum value of $f(x)$.

(2) If $f''(c)>0$, then $f(c)$ is a local minimum value of $f(x)$.

Proof (1) It is tempting to say that, since $f''(c)<0$, $f(x)$ is concave downward near c and to claim that this proves (1). However, to be sure that $f(x)$ is concave downward in a neighborhood of c, we need $f''(x)<0$ in that neighborhood (not just at c), and nothing in our hypothesis guarantees that. We must be a bit more careful.

By definition and hypothesis, we have
$$f''(x)=\lim_{x\to c}\frac{f'(x)-f'(c)}{x-c}=\lim_{x\to c}\frac{f'(x)-0}{x-c}<0.$$
So we can conclude that there is a (possibly small) interval (α,β) around c where
$$\frac{f'(x)}{x-c}<0,\quad x\neq c.$$
But this inequality implies that $f'(x)>0$ for $\alpha<x<c$ and $f'(x)<0$ for $c<x<\beta$. Thus, by the First Derivative Test, $f(c)$ is a local maximum value.

The proofs of (2) is similar.

Example 3 For the function
$$f(x)=x^2-6x+5,$$
use the Second Derivative Test to identify local extrema.

Solution This is the function of Example 5. Note that
$$f'(x)=2x-6=2(x-3),$$
$$f''(x)=2.$$
Thus, $f'(3)=0$ and $f''(3)>0$. Therefore, by the Second Derivative Test, $f(3)$ is a local minimum value.

Example 4 For the function
$$f(x)=\frac{1}{3}x^3-x^2-3x+4,$$
use the Second Derivative Test to identify local extrema.

Solution This is the function of Example 6. We have
$$f'(x)=x^2-2x-3=(x+1)(x-3),$$
$$f''(x)=2x-2.$$
The critical points are -1 and 3 ($f'(-1)=f'(3)=0$). Since $f''(-1)=-4$ and $f''(3)=4$, we conclude by the

(1) 如果 $f''(c)<0$，则 $f(c)$ 为 $f(x)$ 的局部极大值；

(2) 如果 $f''(c)>0$，则 $f(c)$ 为 $f(x)$ 的局部极小值.

证明 (1) 由于 $f''(c)<0$，这倾向于说 $f(x)$ 在 c 附近为下凹函数，从而可断言(1)已得到证明. 但是，为了确定 $f(x)$ 在 c 附近为下凹函数，我们需要在 c 附近也有 $f''(x)<0$(不只是在点 c). 可是，在我们的假设中没有哪一点可以保证那是成立的，我们必须做更仔细的分析.

根据定义和假设，有
$$f''(c)=\lim_{x\to c}\frac{f'(x)-f'(c)}{x-c}=\lim_{x\to c}\frac{f'(x)-0}{x-c}<0.$$
所以可以得出结论：在 c 附近有一个(尽可能小的)区间 (α,β)，在此区间上有
$$\frac{f'(x)}{x-c}<0,\quad x\neq c.$$
而这个不等式意味着，在 $\alpha<x<c$ 时有 $f'(x)>0$，且在 $c<x<\beta$ 时有 $f'(x)<0$. 因此，根据一阶导数判别法，$f(c)$ 是局部极大值.

类似地，可证明(2).

例3 对函数
$$f(x)=x^2-6x+5,$$
用二阶导数判别法求局部极值.

解 这是例5的函数. 注意到
$$f'(x)=2x-6=2(x-3),$$
$$f''(x)=2.$$
所以 $f'(3)=0$ 且 $f''(3)>0$. 因此，根据二阶导数判别法，$f(3)$ 是局部极小值.

例4 对函数
$$f(x)=\frac{1}{3}x^3-x^2-3x+4,$$
用二阶导数法则求局部极值.

解 这是例6的函数. 我们有
$$f'(x)=x^2-2x-3=(x+1)(x-3),$$
$$f''(x)=2x-2.$$
临界点是 -1 和 3 ($f'(-1)=f'(3)=0$). 因为 $f''(-1)=-4$ 和 $f''(3)=4$，所以我们可以

Second Derivative Test that $f(-1)$ is a local maximum value and that $f(3)$ is a local minimum value.

Unfortunately, the Second Derivative Test sometimes fails, since $f''(x)$ may be 0 at a stationary point. For instance, for both $f(x)=x^3$ and $f(x)=x^4$, $f'(0)=0$ and $f''(0)=0$ (Figure 4.27). The first does not have a local maximum or minimum value at 0; the second has a local minimum there. This shows that if $f''(x)=0$ at a stationary point, we are unable to draw a conclusion about maxima or minima without more information.

通过二阶导数判别法得出结论：$f(-1)$是局部极大值，$f(3)$是局部极小值。

遗憾的是，有时二阶导数判别法可能失效，因为$f''(x)$在驻点可能为0. 例如，对于$f(x)=x^3$ 和 $f(x)=x^4$，都有 $f'(0)=0$ 和 $f''(0)=0$（图4.27）。在0处，第一个函数没有局部极值，第二个函数有局部极小值. 这表明，如果仅知道在驻点$f''(x)=0$而没有其他条件，我们不能得出有关极大和极小值的结论.

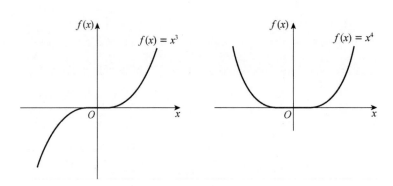

Figure 4.27

图 4.27

2. Extrema on an Open Interval

The problems studied in this section and Section 4.4 are in general under the assumption that the interval on which we wanted to maximize or minimize a function was a closed interval. However, the intervals that arise in practice are not always closed. They are sometimes open, or even open on one end and closed on the other. We can still handle these problems if we correctly apply the theory developed in this section. Keep in mind that maximum (minimum) with no qualifying adjective means global maximum (minimum).

Example 5　Find (if any exist) the minimum and maximum values of the function
$$f(x)=x^4-4x$$
on $(-\infty,+\infty)$.

Solution　The derivative of $f(x)$ is

2. 开区间上的最值

在本节和4.4节所讨论的问题中，通常假设，要求最大值或最小值的函数所在的区间是闭区间. 然而，实际问题中出现的区间往往不都是闭区间. 它们有时是开的，甚至是半开半闭的. 如果能正确应用本节的理论，我们仍能处理这些问题. 记住没有限制的最大（最小）值指的是全局最大（最小）值.

例 5　求函数
$$f(x)=x^4-4x$$
在$(-\infty,+\infty)$上的最大值和最小值（如果存在的话）.

解　$f(x)$的导数为

$$f'(x) = 4x^3 - 4 = 4(x^3 - 1)$$
$$= 4(x-1)(x^2 + x + 1).$$

Since $x^2 + x + 1 = 0$ has no real solution (quadratic formula), there is only one critical point, $x = 1$. For $x < 1$, $f'(x) < 0$, whereas for $x > 1$, $f'(x) > 0$. We conclude that $f(1) = -3$ is a local minimum value of $f(x)$; and since $f(x)$ is decreasing on the left of $x = 1$ and increasing on the right of $x = 1$, it must actually be the minimum value of $f(x)$.

The facts stated above imply that $f(x)$ cannot have a maximum value. The graph of the function $f(x)$ is shown in Figure 4.28.

$$f'(x) = 4x^3 - 4 = 4(x^3 - 1)$$
$$= 4(x-1)(x^2 + x + 1).$$

因为 $x^2 + x + 1 = 0$ 没有实根(二次方程公式), 所以只有一个临界点 $x = 1$. 对 $x < 1$, 有 $f'(x) < 0$; 而对 $x > 1$, 有 $f'(x) > 0$. 由此得出结论: $f(1) = -3$ 是局部极小值. 又因为 $f(x)$ 在 $x = 1$ 的左侧单调递减, 在 $x = 1$ 的右侧单调递增, 所以 $f(1)$ 一定就是 $f(x)$ 的最小值.

上述事实指出, $f(x)$ 不可能有最大值. 函数 $f(x)$ 的图像如图 4.28 所示.

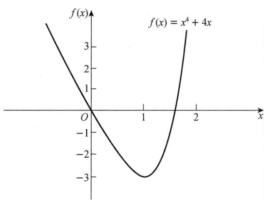

Figure 4.28
图 4.28

4.6 Graphing Functions
4.6 作函数的图像

We propose plotting enough points so that the essential features of the graph are clear, but if the function to be graphed is complicated or if we want a very accurate graph, we must use a powerful tool by calculus providing for analyzing the fine structure of a graph, especially in identifying those points where the character of the graph changes. For example, the local maximum points, the local minimum points, and the inflection points are located, where the graph is increasing or where it is concave up are determined precisely.

In this section, we will introduce the graphing procedure for a function.

我们依据足够多的点就能使图像的基本特点变得明显, 但如果函数很复杂或者需要绘制精确的图像, 我们就必须利用微积分提供的强有力工具来分析图像的细微结构, 特别是在确定图像特征发生改变的关键点方面, 如找到局部极大值点、局部极小值点和拐点, 确定哪些区间上函数图像是单调递增或单调递减的, 是上凹或下凹的.

这一节, 我们将介绍函数绘图的步骤.

Let's illustrate with examples.

Example 1 Sketch the graph of the function
$$f(x)=x^3-3x^2+4.$$

Solution Both the domain and the range of the function $f(x)$ are $(-\infty,+\infty)$. The x-intercepts are -1 and 2, and y-intercepts is 4.

Differentiation twice give
$$f'(x)=3x^2-6x=3x(x-2),$$
$$f''(x)=6x-6=6(x-1).$$

Solving $f'(x)=0$ and $f''(x)=0$ yield
$$x=0,\ x=2,\quad\text{and}\quad x=1.$$

Since $f'(x)>0$ holds on $(-\infty,0)\cup(2,+\infty)$, and $f'(x)<0$ holds on $(0,2)$, $f(x)$ is increasing on $(-\infty,0)\cup(2,+\infty)$ and decreasing on $(0,2)$. Since $f''(x)<0$ for all x in $(-\infty,1)$, and $f''(x)>0$ for all x in $(1,+\infty)$, $f(x)$ is concave down on $(-\infty,1)$ and concave up on $(1,+\infty)$. So the point $(1,2)$ is the inflection point, the points $(0,4)$ and $(2,0)$ are the local maximum point and the local minimum point respectively.

Obviously, there is no asymptotes.

We show the graph of the function $f(x)$ in Figure 4.29.

例 1 画出函数 $f(x)=x^3-3x^2+4$ 的图形.

解 函数 $f(x)$ 的定义域和值域均是 $(-\infty,+\infty)$. $f(x)$ 的图像在 x 轴上的截距为 -1 和 2,在 y 轴上的截距为 4.

对函数求一、二阶导数,得
$$f'(x)=3x^2-6x=3x(x-2),$$
$$f''(x)=6x-6=6(x-1).$$

解 $f'(x)=0$ 和 $f''(x)=0$,得
$$x=0,\ x=2\quad\text{和}\quad x=1.$$

由于在 $(-\infty,+0)\cup(2,+\infty)$ 内有 $f'(x)>0$,在 $(0,2)$ 内有 $f'(x)<0$,所以 $f(x)$ 在 $(-\infty,0)\cup(2,+\infty)$ 内是单调递增的,在 $(0,2)$ 内是单调递减的. 因为在 $(-\infty,1)$ 内有 $f''(x)<0$,在 $(1,+\infty)$ 内有 $f''(x)>0$,所以 $f(x)$ 在 $(-\infty,1)$ 内是下凹的,在 $(1,+\infty)$ 内是上凹的. 故点 $(1,2)$ 是拐点,点 $(0,4)$ 和 $(2,0)$ 分别是局部极大值点和局部极小值点.

显然,不存在渐近线.

函数 $f(x)$ 的图像如图 4.29 所示.

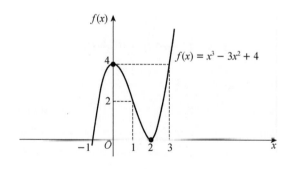

Figure 4.29
图 4.29

Example 2 Analyze the function
$$f(x)=\frac{\sqrt{x}(x-5)^2}{4}.$$
Sketch its graph.

Solution Since the domain of $f(x)$ is $[0,+\infty)$, and the range is $[0,+\infty)$, the graph of $f(x)$ is confined to

例 2 分析函数
$$f(x)=\frac{\sqrt{x}(x-5)^2}{4},$$
画出该函数的图形.

解 由于 $f(x)$ 的定义域是 $[0,+\infty)$,值域是 $[0,+\infty)$, $f(x)$ 的图像被限制

the first quadrant and the positive coordinate axes. The x-intercepts are 0 and 5, and y-intercepts is 0. From
$$f'(x) = \frac{5(x-1)(x-5)}{8\sqrt{x}}, \quad x>0,$$
the stationary points are $x=1$ and $x=5$. Because of $f'(x)>0$ on $(0,1)$ and $(5,+\infty)$, and $f'(x)<0$ on $(1,5)$, $x=1$ is a local maximum point and $x=5$ is a local minimum point. On calculating the second derivative, we have
$$f''(x) = \frac{5(3x^2-6x-5)}{16x^{3/2}}, \quad x>0.$$
The equation $3x^2-6x-5=0$ has one solution in $(0,+\infty)$, that is $1+2\sqrt{6}/3$. As $f''(x)<0$ on $(0,1+2\sqrt{6}/3)$, $f''(x)>0$ on $(1+2\sqrt{6}/3,+\infty)$, so point $(1+2\sqrt{6}/3, f(1+2\sqrt{6}/3))$ is an inflection point.

As x grows large, $f(x)$ grows without bound and much faster than any linear function, there are no asymptotes.

The graph is sketched in Figure 4.30.

在第一象限和正坐标轴内. $f(x)$的图像在x轴上的截距为0和5, 在y轴上的截距为0. 由
$$f'(x) = \frac{5(x-1)(x-5)}{8\sqrt{x}}, \quad x>0,$$
知驻点是 $x=1$ 和 $x=5$. 因为在$(0,1)$和$(5,+\infty)$内有 $f'(x)>0$, 在$(1,5)$内有 $f'(x)<0$, 所以 $x=1$ 是局部极大值点, 且 $x=5$ 是局部极小值点. 计算二阶导, 我们有
$$f''(x) = \frac{5(3x^2-6x-5)}{16x^{3/2}}, \quad x>0.$$
方程$3x^2-6x-5=0$在$(0,+\infty)$上有一个解, 是$1+2\sqrt{6}/3$. 因为在$(0,1+2\sqrt{6}/3)$内有 $f''(x)<0$, 在$(1+2\sqrt{6}/3,+\infty)$内有 $f''(x)>0$, 所以点$(1+2\sqrt{6}/3, f(1+2\sqrt{6}/3))$是一个拐点.

当x增大时, $f(x)$也是无限增大的, 增长的速度比任何线性函数的增长速度都快, 所以没有渐近线.

函数$f(x)$的图像如图4.30所示.

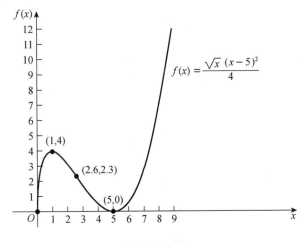

Figure 4.30
图 4.30

Summary of the Method In graphing functions, there is no substitute for common sense. However, the following procedure will be helpful in most cases:

Step 1 Precalculus analysis.

(1) Check the domain and the range of the function to see if any regions of the plane are excluded;

方法总结 绘制函数的图像, 没有一般的方法可套用, 但是以下的步骤在大多数情况下有很大帮助:

第1步 用初等数学方法分析.

(1) 检查函数的定义域和值域, 看看是否能排除平面上的某些区域.

(2) Test for symmetry with respect to the y-axis and the origin;(Is the function even or odd?)

(3) Find the intercepts.

Step 2 Calculus analysis.

(1) Use the first derivative to find the critical points and to find out where the graph is increasing and decreasing.

(2) Test the critical points for the local maxima and the local minima.

(3) Use the second derivative to find out where the graph is concave upward and concave downward and to locate inflection points.

(4) Find the asymptotes.

Step 3 Plot a few points (including all critical points and inflection points).

Step 4 Sketch the graph of the function.

Exercises 4

1. Let the function
$$f(x)=(x-1)(x-2)(x-3)(x-4).$$
How many roots does the equation $f'(x)=0$ have?

2. Let $a>b>0, n>1$. Applying the Lagrange Mean Value Theorem to prove
$$nb^{n-1}(a-b)<a^n-b^n<na^{n-1}(a-b).$$

3. Use the L'Hospital Rule to calculate the following limits:

(1) $\lim\limits_{x\to 1}\dfrac{x^3-3x+2}{x^3-x^2-x+1}$;

(2) $\lim\limits_{x\to\infty}\left(\dfrac{\pi}{2}-\arctan 2x^2\right)x^2$;

(3) $\lim\limits_{x\to 0}\left(\dfrac{1}{x^2}-\cot^2 x\right)$;

(4) $\lim\limits_{x\to 0^+}(\arcsin x)^{\tan x}$;

(5) $\lim\limits_{x\to 0}\left[\dfrac{1}{\ln(1+x)}-\dfrac{1}{x}\right]$;

(6) $\lim\limits_{x\to 0}x\cot 2x$;

(7) $\lim\limits_{x\to 0}\dfrac{x-\sin x}{x^3}$;

(8) $\lim\limits_{x\to 0}\dfrac{\tan x-x}{x\sin x\tan x}$.

4. Find the monotone intervals of the following functions:

(1) $f(x)=e^x-x+1$;

(2) $y=\ln(x+\sqrt{100+x^2})$.

5. Let $f''(x)>0, x\in(-\infty,+\infty), f(0)\leqslant 0$. Find the monotone intervals of the function $\dfrac{f(x)}{x}$.

6. Find the inflection points of the following functions:

(1) $y=x^3$;

(2) $y=x+\dfrac{x}{x^2-1}$;

(3) $y=x^3-5x^2+3x+5$;

(4) $y=x^4(12\ln x-7)$.

7. Prove $\dfrac{e^x+e^y}{2}>e^{(x+y)/2}$ $(x\neq y)$.

8. Find the local extreme values of the following functions:

(1) $f(x)=2x^3-9x^2+12x-3$;

(2) $f(x)=x^{\frac{1}{x}}$;

(3) $y=x+\dfrac{x}{x^2-1}$;

(4) $f(x)=2x-\ln(1+x)$;

(5) $f(x)=(x-4)\sqrt[3]{(x+1)^2}$.

9. Find the extreme values of the function
$$f(x)=2x^3-6x^2-18x-7$$
on the interval $[1,4]$.

10. Find the maximum value of the function
$$f(x)=\dfrac{x}{x^2+1}\quad(x\geqslant 0).$$

11. Sketch the graphs of the following functions:

(1) $f(x)=\dfrac{x^2}{2x-1}$;

(2) $f(x)=x^4-2x^3+1$;

(3) $f(x)=x+|\sin 2x|$.

Chapter 5　The Indefinite Integral
第 5 章　不定积分

5.1　The Concept and the Properties of the Indefinite Integral
5.1　不定积分的概念与性质

1. The Concepts of the Primitive Function and the Indefinite Integral

Definition 5.1　A function $F(x)$ is called the **primitive function** of $f(x)$ on the interval I if the derivative of differentiable function $F(x)$ is $f(x)$, that is, for any $x \in I$,
$$F'(x) = f(x).$$

For example, since $(\sin x)' = \cos x$, $\sin x$ is the primitive function of $\cos x$. And for $x \in (0, +\infty)$, since
$$(\sqrt{x})' = \frac{1}{2\sqrt{x}},$$
$\sqrt{x}$ is the primitive function of $\frac{1}{2\sqrt{x}}$ on $(0, +\infty)$.

Existence Theorem for the Primitive Function　If $f(x)$ is continuous on the interval I, then there exists a differentiable function $F(x)$, for any $x \in I$,
$$F'(x) = f(x).$$

In short, continuous functions have primitive function certainly.

If $F(x)$ is the primitive function of $f(x)$ on the interval I, then $f(x)$ have infinite primitive functions, $F(x) + C$ are all the primitive functions of $f(x)$, where C is an arbitrary constant.

The difference of any two primitive functions of $f(x)$ is a constant, or if $\Phi(x)$ and $F(x)$ are the primitive functions of $f(x)$, then
$$\Phi(x) - F(x) = C,$$
where C is a constant.

1. 原函数与不定积分的概念

定义 5.1　如果在区间 I 上，可导函数 $F(x)$ 的导函数为 $f(x)$，即对任一 $x \in I$，有
$$F'(x) = f(x),$$
那么函数 $F(x)$ 称为 $f(x)$ 在区间 I 上的**原函数**。

例如，因为 $(\sin x)' = \cos x$，所以 $\sin x$ 是 $\cos x$ 的原函数。又如，当 $x \in (0, +\infty)$ 时，因为
$$(\sqrt{x})' = \frac{1}{2\sqrt{x}},$$
所以 $\sqrt{x}$ 是 $\frac{1}{2\sqrt{x}}$ 在 $(0, +\infty)$ 上的原函数。

原函数存在定理　如果函数 $f(x)$ 在区间 I 上连续，那么在区间 I 上存在可导函数 $F(x)$，使得对任一 $x \in I$，有
$$F'(x) = f(x).$$

简单地说，连续函数一定有原函数。

如果函数 $f(x)$ 在区间 I 上有原函数 $F(x)$，那么 $f(x)$ 就有无限多个原函数，$F(x) + C$ 都是 $f(x)$ 的原函数，其中 C 是任意常数。

函数 $f(x)$ 的任意两个原函数之间只差一个常数，即如果 $\Phi(x)$ 和 $F(x)$ 都是 $f(x)$ 的原函数，则
$$\Phi(x) - F(x) = C,$$
其中 C 为常数。

Definition 5.2 On the interval I, a primitive function of $f(x)$ with any constant is called the **indefinite integral** of $f(x)$, writing as
$$\int f(x)\mathrm{d}x.$$

The symbol $\int$ is called the **sign of integration**, $f(x)$ is called the **integrand function**, $f(x)\mathrm{d}x$ is called the **integral expression**, and x is called the **integration variable**.

According to the definition, if $F(x)$ is a primitive function of $f(x)$ on interval I, then the indefinite integral of $f(x)$ is $F(x)+C$, that is
$$\int f(x)\mathrm{d}x = F(x)+C.$$

So $\int f(x)\mathrm{d}x$ can be expressed as any primitive function of $f(x)$.

Example 1 $\sin x$ is a primitive function of $\cos x$, so
$$\int \cos x \mathrm{d}x = \sin x + C.$$

$\sqrt{x}$ is the primitive function of $\dfrac{1}{2\sqrt{x}}$, so
$$\int \frac{1}{2\sqrt{x}}\mathrm{d}x = \sqrt{x}+C.$$

Example 2 Compute the indefinite integral of $f(x)=\dfrac{1}{x}$.

Solution If $x>0$, $(\ln x)'=\dfrac{1}{x}$, so
$$\int \frac{1}{x}\mathrm{d}x = \ln x + C \quad (x>0).$$

If $x<0$, $[\ln(-x)]'=\dfrac{1}{-x}\cdot(-1)=\dfrac{1}{x}$, so
$$\int \frac{1}{x}\mathrm{d}x = \ln(-x) + C \quad (x<0).$$

In summary, we have
$$\int \frac{1}{x}\mathrm{d}x = \ln|x| + C \quad (x\neq 0).$$

Example 3 Suppose that a curve passes through the point $(1,2)$, and the slope of tangent of any point on the curve is equal to the double abscissa. Compute the function of the curve.

定义 5.2 在区间 I 上,函数 $f(x)$ 的带有任意常数的原函数称为 $f(x)$ 在区间 I 上的**不定积分**,记作
$$\int f(x)\mathrm{d}x,$$

其中记号 $\int$ 称为**积分号**,$f(x)$ 称为**被积函数**,$f(x)\mathrm{d}x$ 称为**被积表达式**,x 称为**积分变量**.

根据定义,如果 $F(x)$ 是 $f(x)$ 在区间 I 上的一个原函数,那么 $F(x)+C$ 就是 $f(x)$ 的不定积分,即
$$\int f(x)\mathrm{d}x = F(x)+C.$$

因而不定积分 $\int f(x)\mathrm{d}x$ 可以表示 $f(x)$ 的任意一个原函数.

例 1 因为 $\sin x$ 是 $\cos x$ 的原函数,所以
$$\int \cos x \mathrm{d}x = \sin x + C.$$

因为 $\sqrt{x}$ 是 $\dfrac{1}{2\sqrt{x}}$ 的原函数,所以
$$\int \frac{1}{2\sqrt{x}}\mathrm{d}x = \sqrt{x}+C.$$

例 2 求 $f(x)=\dfrac{1}{x}$ 的不定积分.

解 当 $x>0$ 时,$(\ln x)'=\dfrac{1}{x}$,所以
$$\int \frac{1}{x}\mathrm{d}x = \ln x + C \quad (x>0);$$

当 $x<0$ 时,$[\ln(-x)]'=\dfrac{1}{-x}\cdot(-1)=\dfrac{1}{x}$,所以
$$\int \frac{1}{x}\mathrm{d}x = \ln(-x) + C \quad (x<0).$$

综上所述,得到
$$\int \frac{1}{x}\mathrm{d}x = \ln|x| + C \quad (x\neq 0).$$

例 3 设曲线通过点 $(1,2)$,且其上任一点处的切线斜率等于这点横坐标的两倍,求此曲线的方程.

Solution Suppose that the function is $y=f(x)$. According to the problem, the slope of tangent of any point (x,y) on the curve is
$$y'=f'(x)=2x,$$
then $f(x)$ is a primitive function of $2x$.

Since $\int 2x\mathrm{d}x=x^2+C$, then exists a constant C_0, so that $f(x)=x^2+C_0$, or the function of the curve is
$$y=x^2+C_0.$$
And the curve passes through the point $(1,2)$, so $2=1+C_0$, $C_0=1$, then the function of the curve is
$$y=x^2+1.$$

The graph of the primitive functions of the function $f(x)$ is called the **integral curves** of $f(x)$. In Example 3, the curve $y=x^2+1$ is the integral curve of $2x$, which passes through the point $(1,2)$.

According to the definition of indefinite integral, we know:
$$\frac{\mathrm{d}}{\mathrm{d}x}\left[\int f(x)\mathrm{d}x\right]=f(x),$$
or
$$\mathrm{d}\left[\int f(x)\mathrm{d}x\right]=f(x)\mathrm{d}x;$$
And $F(x)$ is a primitive function of $F'(x)$, then
$$\int F'(x)\mathrm{d}x=F(x)+C,$$
or
$$\int \mathrm{d}F(x)=F(x)+C.$$

It follows that, differential operation (shows as symbol d) and indefinite integral operation (short as integral operation, shows as symbol $\int$) are mutually inverse. When the symbol $\int$ interaction with d, they are counteract, or counteract with a constant.

2. Basic Formulas of Integrals

(1) $\int k\mathrm{d}x=kx+C$ (k is a constant);

(2) $\int x^\mu \mathrm{d}x = \dfrac{1}{\mu+1} x^{\mu+1} + C$ ($\mu \neq -1$);

(3) $\int \dfrac{1}{x} \mathrm{d}x = \ln|x| + C$;

(4) $\int \mathrm{e}^x \mathrm{d}x = \mathrm{e}^x + C$;

(5) $\int a^x \mathrm{d}x = \dfrac{a^x}{\ln a} + C$ ($a > 0$, $a \neq 1$);

(6) $\int \cos x \mathrm{d}x = \sin x + C$;

(7) $\int \sin x \mathrm{d}x = -\cos x + C$;

(8) $\int \dfrac{1}{\cos^2 x} \mathrm{d}x = \int \sec^2 x \mathrm{d}x = \tan x + C$;

(9) $\int \dfrac{1}{\sin^2 x} \mathrm{d}x = \int \csc^2 x \mathrm{d}x = -\cot x + C$;

(10) $\int \dfrac{1}{1+x^2} \mathrm{d}x = \arctan x + C$;

(11) $\int \dfrac{1}{\sqrt{1-x^2}} \mathrm{d}x = \arcsin x + C$;

(12) $\int \sec x \tan x \mathrm{d}x = \sec x + C$;

(13) $\int \csc x \cot x \mathrm{d}x = -\csc x + C$;

Example 4 Compute $\int \dfrac{1}{x^3} \mathrm{d}x$.

Solution
$$\int \dfrac{1}{x^3} \mathrm{d}x = \int x^{-3} \mathrm{d}x = \dfrac{1}{-3+1} x^{-3+1} + C$$
$$= -\dfrac{1}{2x^2} + C.$$

Example 5 Compute $\int x^2 \sqrt{x} \mathrm{d}x$.

Solution
$$\int x^2 \sqrt{x} \mathrm{d}x = \int x^{\frac{5}{2}} \mathrm{d}x = \dfrac{1}{\frac{5}{2}+1} x^{\frac{5}{2}+1} + C$$
$$= \dfrac{2}{7} x^{\frac{7}{2}} + C = \dfrac{2}{7} x^3 \sqrt{x} + C.$$

Example 6 Compute $\int \dfrac{1}{x^3 \sqrt{x}} \mathrm{d}x$.

Solution
$$\int \dfrac{1}{x^3 \sqrt{x}} \mathrm{d}x = \int x^{-\frac{7}{2}} \mathrm{d}x = \dfrac{x^{-\frac{7}{2}+1}}{-\frac{7}{2}+1} + C$$
$$= -\dfrac{2}{5} x^{-\frac{5}{2}} + C = -\dfrac{2}{5\sqrt{x^5}} + C.$$

3. The Properties of the Indefinite Integral

Property 1 The indefinite integral of the sum of two functions is equal to the sum of the indefinite integrals of every function, that is
$$\int [f(x)+g(x)]dx = \int f(x)dx + \int g(x)dx.$$

This is because of
$$\left[\int f(x)dx + \int g(x)dx\right]'$$
$$= \left[\int f(x)dx\right]' + \left[\int g(x)dx\right]'$$
$$= f(x) + g(x).$$

Property 2 In the operation of indefinite integrals, the nonzero constant in the integrand function could be move to the out of the symbol $\int$, that is
$$\int kf(x)dx = k\int f(x)dx \quad (k \text{ is a nonzero constant}).$$

Example 7 Compute $\int \sqrt{x}(x^2-5)dx$.

Solution
$$\int \sqrt{x}(x^2-5)dx$$
$$= \int (x^{\frac{5}{2}} - 5x^{\frac{1}{2}})dx$$
$$= \int x^{\frac{5}{2}}dx - \int 5x^{\frac{1}{2}}dx$$
$$= \int x^{\frac{5}{2}}dx - 5\int x^{\frac{1}{2}}dx$$
$$= \frac{2}{7}x^{\frac{7}{2}} - 5 \cdot \frac{2}{3}x^{\frac{3}{2}} + C.$$

Example 8 Compute $\int \frac{(x-1)^3}{x^2}dx$.

Solution
$$\int \frac{(x-1)^3}{x^2}dx$$
$$= \int \frac{x^3-3x^2+3x-1}{x^2}dx$$
$$= \int \left(x-3+\frac{3}{x}-\frac{1}{x^2}\right)dx$$
$$= \int xdx - 3\int dx + 3\int \frac{1}{x}dx - \int \frac{1}{x^2}dx$$
$$= \frac{1}{2}x^2 - 3x + 3\ln|x| + \frac{1}{x} + C.$$

Example 9 Compute $\int (e^x - 3\cos x)dx$.

3. 不定积分的性质

性质 1 两个函数的和的不定积分等于两个函数的不定积分的和，即
$$\int [f(x)+g(x)]dx = \int f(x)dx + \int g(x)dx.$$

这是因为
$$\left[\int f(x)dx + \int g(x)dx\right]'$$
$$= \left[\int f(x)dx\right]' + \left[\int g(x)dx\right]'$$
$$= f(x) + g(x).$$

性质 2 在不定积分运算中，被积函数中的非零常数因子可以提到积分号 $\int$ 外面来，即
$$\int kf(x)dx = k\int f(x)dx \quad (k \text{ 是非零常数}).$$

例 7 求 $\int \sqrt{x}(x^2-5)dx$.

解
$$\int \sqrt{x}(x^2-5)dx$$
$$= \int (x^{\frac{5}{2}} - 5x^{\frac{1}{2}})dx$$
$$= \int x^{\frac{5}{2}}dx - \int 5x^{\frac{1}{2}}dx$$
$$= \int x^{\frac{5}{2}}dx - 5\int x^{\frac{1}{2}}dx$$
$$= \frac{2}{7}x^{\frac{7}{2}} - 5 \cdot \frac{2}{3}x^{\frac{3}{2}} + C.$$

例 8 求 $\int \frac{(x-1)^3}{x^2}dx$.

解
$$\int \frac{(x-1)^3}{x^2}dx$$
$$= \int \frac{x^3-3x^2+3x-1}{x^2}dx$$
$$= \int \left(x-3+\frac{3}{x}-\frac{1}{x^2}\right)dx$$
$$= \int xdx - 3\int dx + 3\int \frac{1}{x}dx - \int \frac{1}{x^2}dx$$
$$= \frac{1}{2}x^2 - 3x + 3\ln|x| + \frac{1}{x} + C.$$

例 9 求 $\int (e^x - 3\cos x)dx$.

Solution $\int (e^x - 3\cos x) dx$

$= \int e^x dx - 3\int \cos x dx$

$= e^x - 3\sin x + C.$

Example 10 Compute $\int 2^x e^x dx$.

Solution $\int 2^x e^x dx = \int (2e)^x dx$

$= \dfrac{(2e)^x}{\ln(2e)} + C$

$= \dfrac{2^x e^x}{1 + \ln 2} + C.$

Example 11 Compute $\int \tan^2 x dx$.

Solution $\int \tan^2 x dx = \int (\sec^2 x - 1) dx$

$= \int \sec^2 x dx - \int dx$

$= \tan x - x + C.$

Example 12 Compute $\int \sin^2 \dfrac{x}{2} dx$.

Solution $\int \sin^2 \dfrac{x}{2} dx = \int \dfrac{1 - \cos x}{2} dx$

$= \dfrac{1}{2} \int (1 - \cos x) dx$

$= \dfrac{1}{2}(x - \sin x) + C.$

Example 13 Compute $\int \dfrac{1}{\sin^2 \dfrac{x}{2} \cos^2 \dfrac{x}{2}} dx$.

Solution $\int \dfrac{1}{\sin^2 \dfrac{x}{2} \cos^2 \dfrac{x}{2}} dx = 4\int \dfrac{1}{\sin^2 x} dx$

$= -4\cot x + C.$

Example 14 Compute $\int \dfrac{2x^4 + x^2 + 3}{x^2 + 1} dx$.

Solution $\int \dfrac{2x^4 + x^2 + 3}{x^2 + 1} dx$

$= \int \left(2x^2 - 1 + \dfrac{4}{x^2 + 1}\right) dx$

$= 2\int x^2 dx - \int dx + 4\int \dfrac{1}{x^2 + 1} dx$

$= \dfrac{2}{3} x^3 - x + 4\arctan x + C.$

5.2 Integration by Substitution

1. The Substitution Rule 1

Theorem 5.1 (The Substitution Rule 1) If $u=\varphi(x)$ is a differentiable function whose range is an interval I, and $f(u)$ is continuous on I, then

$$\int f(\varphi(x))\varphi'(x)\mathrm{d}x = \left[\int f(u)\mathrm{d}u\right]_{u=\varphi(x)}$$

Proof By the Chain Rule, $F(\varphi(x))$ is a primitive function of $f(\varphi(x))\varphi'(x)$, where F is a primitive function of f:

$$\frac{\mathrm{d}}{\mathrm{d}x}F(\varphi(x)) = F'(\varphi(x))\varphi'(x)$$
$$= f(\varphi(x))\varphi'(x).$$

If we make the substitution $u=\varphi(x)$, then

$$\int f(\varphi(x))\varphi'(x)\mathrm{d}x = \int \frac{\mathrm{d}}{\mathrm{d}x}F((\varphi(x))\mathrm{d}x$$
$$= F(\varphi(x)) + C = F(u) + C$$
$$= \int F'(u)\mathrm{d}u = \int f(u)\mathrm{d}u$$
$$= \left[\int f(u)\mathrm{d}u\right]_{u=\varphi(x)}.$$

The Substitution Rule 1 provides the following substitution method to evaluate the indefinite integral $\int f(\varphi(x))\varphi'(x)\mathrm{d}x$:

(1) Substitute $u=\varphi(x)$ and $\mathrm{d}u=\frac{\mathrm{d}u}{\mathrm{d}x}\mathrm{d}x = \varphi'(x)\mathrm{d}x$ to obtain the indefinite integral $\int f(u)\mathrm{d}u$.

(2) Replace u by $\varphi(x)$ in the result.

Example 1 Compute $\int 2\cos 2x\,\mathrm{d}x$.

Solution
$$\int 2\cos 2x\,\mathrm{d}x = \int \cos 2x \cdot (2x)'\mathrm{d}x$$
$$= \int \cos 2x\,\mathrm{d}(2x)$$
$$= \int \cos u\,\mathrm{d}u = \sin u + C$$
$$= \sin 2x + C.$$

1. 第一换元法

定理 5.1(第一换元法) 如果 $u=\varphi(x)$ 是可导函数,其值域是区间 I,$f(u)$ 在区间 I 上连续,那么

$$\int f(\varphi(x))\varphi'(x)\mathrm{d}x = \left[\int f(u)\mathrm{d}u\right]_{u=\varphi(x)}$$

证明 根据链式法则,$F(\varphi(x))$ 是 $f(\varphi(x))\varphi'(x)$ 的一个原函数,这里 F 是 f 的一个原函数:

$$\frac{\mathrm{d}}{\mathrm{d}x}F(\varphi(x)) = F'(\varphi(x))\varphi'(x)$$
$$= f(\varphi(x))\varphi'(x).$$

若我们作替换 $u=\varphi(x)$,那么

$$\int f(\varphi(x))\varphi'(x)\mathrm{d}x = \int \frac{\mathrm{d}}{\mathrm{d}x}F((\varphi(x))\mathrm{d}x$$
$$= F[\varphi(x)] + C = F(u) + C$$
$$= \int F'(u)\mathrm{d}u = \int f(u)\mathrm{d}u$$
$$= \left[\int f(u)\mathrm{d}u\right]_{u=\varphi(x)}.$$

第一换元法为求解形如 $\int f(\varphi(x))\varphi'(x)\mathrm{d}x$ 的不定积分提供了如下的替换方法:

(1) 利用替换 $u=\varphi(x)$ 和 $\mathrm{d}u=\frac{\mathrm{d}u}{\mathrm{d}x}\mathrm{d}x = \varphi'(x)\mathrm{d}x$ 来求解不定积分 $\int f(u)\mathrm{d}u$;

(2) 在结果中用 $\varphi(x)$ 替代 u.

例 1 求 $\int 2\cos 2x\,\mathrm{d}x$.

解
$$\int 2\cos 2x\,\mathrm{d}x = \int \cos 2x \cdot (2x)'\mathrm{d}x$$
$$= \int \cos 2x\,\mathrm{d}(2x)$$
$$= \int \cos u\,\mathrm{d}u = \sin u + C$$
$$= \sin 2x + C.$$

Example 2 Compute $\int \dfrac{1}{3+2x}dx$.

Solution
$$\int \dfrac{1}{3+2x}dx = \dfrac{1}{2}\int \dfrac{1}{3+2x}(3+2x)'dx$$
$$= \dfrac{1}{2}\int \dfrac{1}{3+2x}d(3+2x)$$
$$\xlongequal{\text{Let } u=3+2x} \dfrac{1}{2}\int \dfrac{1}{u}du$$
$$= \dfrac{1}{2}\ln|u|+C$$
$$= \dfrac{1}{2}\ln|3+2x|+C.$$

Corollary 1
$$\int f(ax+b)dx$$
$$= \int \dfrac{1}{a}f(ax+b)d(ax+b)$$
$$= \dfrac{1}{a}\left[\int f(u)du\right]_{u=ax+b}.$$

Example 3 Compute $\int \dfrac{x^2}{(x+2)^3}dx$.

Solution Let $u=x+2$, $x=u-2$, then $dx=du$, so
$$\int \dfrac{x^2}{(x+2)^3}dx = \int \dfrac{(u-2)^2}{u^3}du$$
$$= \int (u^2-4u+4)u^{-3}du$$
$$= \int (u^{-1}-4u^{-2}+4u^{-3})du$$
$$= \ln|u|+4u^{-1}-2u^{-2}+C$$
$$= \ln|x+2|+\dfrac{4}{x+2}-\dfrac{2}{(x+2)^2}+C.$$

Example 4 Compute $\int 2xe^{x^2}dx$.

Solution
$$\int 2xe^{x^2}dx = \int e^{x^2}(x^2)'dx = \int e^{x^2}d(x^2)$$
$$= \int e^u du = e^u+C = e^{x^2}+C.$$

Example 5 Compute $\int x\sqrt{1-x^2}\,dx$.

Solution
$$\int x\sqrt{1-x^2}\,dx = \dfrac{1}{2}\int \sqrt{1-x^2}(x^2)'dx$$
$$= \dfrac{1}{2}\int \sqrt{1-x^2}\,d(x^2)$$
$$= -\dfrac{1}{2}\int \sqrt{1-x^2}\,d(1-x^2)$$

例 2 求 $\int \dfrac{1}{3+2x}dx$.

解
$$\int \dfrac{1}{3+2x}dx = \dfrac{1}{2}\int \dfrac{1}{3+2x}(3+2x)'dx$$
$$= \dfrac{1}{2}\int \dfrac{1}{3+2x}d(3+2x)$$
$$\xlongequal{\text{令 } u=3+2x} \dfrac{1}{2}\int \dfrac{1}{u}du$$
$$= \dfrac{1}{2}\ln|u|+C$$
$$= \dfrac{1}{2}\ln|3+2x|+C.$$

推论 1
$$\int f(ax+b)dx$$
$$= \int \dfrac{1}{a}f(ax+b)d(ax+b)$$
$$= \dfrac{1}{a}\left[\int f(u)du\right]_{u=ax+b}.$$

例 3 求 $\int \dfrac{x^2}{(x+2)^3}dx$.

解 令 $u=x+2$, $x=u-2$, 则 $dx=du$. 所以
$$\int \dfrac{x^2}{(x+2)^3}dx = \int \dfrac{(u-2)^2}{u^3}du$$
$$= \int (u^2-4u+4)u^{-3}du$$
$$= \int (u^{-1}-4u^{-2}+4u^{-3})du$$
$$= \ln|u|+4u^{-1}-2u^{-2}+C$$
$$= \ln|x+2|+\dfrac{4}{x+2}-\dfrac{2}{(x+2)^2}+C.$$

例 4 求 $\int 2xe^{x^2}dx$.

解
$$\int 2xe^{x^2}dx = \int e^{x^2}(x^2)'dx = \int e^{x^2}d(x^2)$$
$$= \int e^u du = e^u+C = e^{x^2}+C.$$

例 5 求 $\int x\sqrt{1-x^2}\,dx$.

解
$$\int x\sqrt{1-x^2}\,dx = \dfrac{1}{2}\int \sqrt{1-x^2}(x^2)'dx$$
$$= \dfrac{1}{2}\int \sqrt{1-x^2}\,d(x^2)$$
$$= -\dfrac{1}{2}\int \sqrt{1-x^2}\,d(1-x^2)$$

$$\xrightarrow{\text{Let } u=1-x^2} -\frac{1}{2}\int u^{\frac{1}{2}} du$$

$$= -\frac{1}{3}u^{\frac{3}{2}} + C$$

$$= -\frac{1}{3}(1-x^2)^{\frac{3}{2}} + C.$$

Corollary 2 $\int kxf(ax^2+b)dx$

$$= \int \frac{k}{2a}f(ax^2+b)d(ax^2+b)$$

$$= \frac{k}{2a}\left[\int f(u)du\right]_{u=ax^2+b}$$

Example 6 Compute $\int \frac{1}{a^2+x^2}dx$.

Solution $\int \frac{1}{a^2+x^2}dx = \frac{1}{a^2}\int \frac{1}{1+\left(\frac{x}{a}\right)^2}dx$

$$= \frac{1}{a}\int \frac{1}{1+\left(\frac{x}{a}\right)^2}d\left(\frac{x}{a}\right)$$

$$= \frac{1}{a}\arctan\frac{x}{a} + C.$$

Example 7 Compute $\int \frac{1}{\sqrt{a^2-x^2}}dx \ (a>0)$.

Solution $\int \frac{1}{\sqrt{a^2-x^2}}dx = \frac{1}{a}\int \frac{1}{\sqrt{1-\left(\frac{x}{a}\right)^2}}dx$

$$= \int \frac{1}{\sqrt{1-\left(\frac{x}{a}\right)^2}}d\left(\frac{x}{a}\right)$$

$$= \arcsin\frac{x}{a} + C.$$

Example 8 Compute $\int \frac{1}{x^2-a^2}dx$.

Solution $\int \frac{1}{x^2-a^2}dx = \frac{1}{2a}\int \left(\frac{1}{x-a} - \frac{1}{x+a}\right)dx$

$$= \frac{1}{2a}\left(\int \frac{1}{x-a}dx - \int \frac{1}{x+a}dx\right)$$

$$= \frac{1}{2a}\left[\int \frac{1}{x-a}d(x-a) - \int \frac{1}{x+a}d(x+a)\right]$$

$$= \frac{1}{2a}(\ln|x-a| - \ln|x+a|) + C$$

$$= \frac{1}{2a}\ln\left|\frac{x-a}{x+a}\right| + C.$$

Example 9 Compute $\int \dfrac{1}{x(1+2\ln x)}\mathrm{d}x$.

Solution $\int \dfrac{1}{x(1+2\ln x)}\mathrm{d}x = \int \dfrac{\mathrm{d}(\ln x)}{1+2\ln x}$

$$= \dfrac{1}{2}\int \dfrac{\mathrm{d}(1+2\ln x)}{1+2\ln x}$$

$$= \dfrac{1}{2}\ln|1+2\ln x| + C.$$

Example 10 Compute $\int \dfrac{\mathrm{e}^{3\sqrt{x}}}{\sqrt{x}}\mathrm{d}x$.

Solution $\int \dfrac{\mathrm{e}^{3\sqrt{x}}}{\sqrt{x}}\mathrm{d}x = 2\int \mathrm{e}^{3\sqrt{x}}\mathrm{d}(\sqrt{x})$

$$= \dfrac{2}{3}\int \mathrm{e}^{3\sqrt{x}}\mathrm{d}(3\sqrt{x})$$

$$= \dfrac{2}{3}\mathrm{e}^{3\sqrt{x}} + C.$$

Example 11 Compute $\int \sin^3 x \mathrm{d}x$.

Solution $\int \sin^3 x \mathrm{d}x = \int \sin^2 x \cdot \sin x \mathrm{d}x$

$$= -\int (1-\cos^2 x)\mathrm{d}(\cos x)$$

$$= -\int \mathrm{d}(\cos x) + \int \cos^2 x \mathrm{d}(\cos x)$$

$$= -\cos x + \dfrac{1}{3}\cos^3 x + C.$$

Example 12 Compute $\int \sin^2 x \cos^5 x \mathrm{d}x$.

Solution $\int \sin^2 x \cos^5 x \mathrm{d}x = \int \sin^2 x \cos^4 x \mathrm{d}(\sin x)$

$$= \int \sin^2 x (1-\sin^2 x)^2 \mathrm{d}(\sin x)$$

$$= \int (\sin^2 x - 2\sin^4 x + \sin^6 x)\mathrm{d}(\sin x)$$

$$= \dfrac{1}{3}\sin^3 x - \dfrac{2}{5}\sin^5 x + \dfrac{1}{7}\sin^7 x + C.$$

Example 13 Compute $\int \tan x \mathrm{d}x$.

Solution $\int \tan x \mathrm{d}x = \int \dfrac{\sin x}{\cos x}\mathrm{d}x$

$$= -\int \dfrac{1}{\cos x}\mathrm{d}(\cos x)$$

$$= -\ln|\cos x| + C.$$

例 9 求 $\int \dfrac{1}{x(1+2\ln x)}\mathrm{d}x$.

解 $\int \dfrac{1}{x(1+2\ln x)}\mathrm{d}x = \int \dfrac{\mathrm{d}(\ln x)}{1+2\ln x}$

$$= \dfrac{1}{2}\int \dfrac{\mathrm{d}(1+2\ln x)}{1+2\ln x}$$

$$= \dfrac{1}{2}\ln|1+2\ln x| + C.$$

例 10 求 $\int \dfrac{\mathrm{e}^{3\sqrt{x}}}{\sqrt{x}}\mathrm{d}x$.

解 $\int \dfrac{\mathrm{e}^{3\sqrt{x}}}{\sqrt{x}}\mathrm{d}x = 2\int \mathrm{e}^{3\sqrt{x}}\mathrm{d}(\sqrt{x})$

$$= \dfrac{2}{3}\int \mathrm{e}^{3\sqrt{x}}\mathrm{d}(3\sqrt{x})$$

$$= \dfrac{2}{3}\mathrm{e}^{3\sqrt{x}} + C.$$

例 11 求 $\int \sin^3 x \mathrm{d}x$.

解 $\int \sin^3 x \mathrm{d}x = \int \sin^2 x \cdot \sin x \mathrm{d}x$

$$= -\int (1-\cos^2 x)\mathrm{d}(\cos x)$$

$$= -\int \mathrm{d}(\cos x) + \int \cos^2 x \mathrm{d}(\cos x)$$

$$= -\cos x + \dfrac{1}{3}\cos^3 x + C.$$

例 12 求 $\int \sin^2 x \cos^5 x \mathrm{d}x$.

解 $\int \sin^2 x \cos^5 x \mathrm{d}x = \int \sin^2 x \cos^4 x \mathrm{d}(\sin x)$

$$= \int \sin^2 x (1-\sin^2 x)^2 \mathrm{d}(\sin x)$$

$$= \int (\sin^2 x - 2\sin^4 x + \sin^6 x)\mathrm{d}(\sin x)$$

$$= \dfrac{1}{3}\sin^3 x - \dfrac{2}{5}\sin^5 x + \dfrac{1}{7}\sin^7 x + C.$$

例 13 求 $\int \tan x \mathrm{d}x$.

解 $\int \tan x \mathrm{d}x = \int \dfrac{\sin x}{\cos x}\mathrm{d}x$

$$= -\int \dfrac{1}{\cos x}\mathrm{d}(\cos x)$$

$$= -\ln|\cos x| + C.$$

In the same way, we get
$$\int \cot x\,dx = \ln|\sin x| + C.$$

Example 14 Compute $\int \cos^2 x\,dx$.

Solution
$$\begin{aligned}
\int \cos^2 x\,dx &= \int \frac{1+\cos 2x}{2}\,dx \\
&= \frac{1}{2}\left(\int dx + \int \cos 2x\,dx\right) \\
&= \frac{1}{2}\int dx + \frac{1}{4}\int \cos 2x\,d(2x) \\
&= \frac{1}{2}x + \frac{1}{4}\sin 2x + C.
\end{aligned}$$

Example 15 Compute $\int \cos^4 x\,dx$.

Solution
$$\begin{aligned}
\int \cos^4 x\,dx &= \int (\cos^2 x)^2\,dx \\
&= \int \left[\frac{1}{2}(1+\cos 2x)\right]^2 dx \\
&= \frac{1}{4}\int (1 + 2\cos 2x + \cos^2 2x)\,dx \\
&= \frac{1}{4}\int \left(\frac{3}{2} + 2\cos 2x + \frac{1}{2}\cos 4x\right) dx \\
&= \frac{1}{4}\left(\frac{3}{2}x + \sin 2x + \frac{1}{8}\sin 4x\right) + C \\
&= \frac{3}{8}x + \frac{1}{4}\sin 2x + \frac{1}{32}\sin 4x + C.
\end{aligned}$$

Example 16 Compute $\int \sin^2 x \cos^4 x\,dx$.

Solution
$$\begin{aligned}
\int &\sin^2 x \cos^4 x\,dx \\
&= \frac{1}{8}\int (1-\cos 2x)(1+\cos 2x)^2\,dx \\
&= \frac{1}{8}\int (1 + \cos 2x - \cos^2 2x - \cos^3 2x)\,dx \\
&= \frac{1}{8}\int (\cos 2x - \cos^3 2x)\,dx \\
&\quad + \frac{1}{8}\int (1 - \cos^2 2x)\,dx \\
&= \frac{1}{8}\int \frac{1}{2}\sin^2 2x\,d(\sin 2x) \\
&\quad + \frac{1}{8}\int \frac{1}{2}(1-\cos 4x)\,dx \\
&= \frac{1}{48}\sin^3 2x + \frac{x}{16} - \frac{1}{64}\sin 4x + C.
\end{aligned}$$

同样，可得到
$$\int \cot x\,dx = \ln|\sin x| + C.$$

例 14 求 $\int \cos^2 x\,dx$.

解
$$\begin{aligned}
\int \cos^2 x\,dx &= \int \frac{1+\cos 2x}{2}\,dx \\
&= \frac{1}{2}\left(\int dx + \int \cos 2x\,dx\right) \\
&= \frac{1}{2}\int dx + \frac{1}{4}\int \cos 2x\,d(2x) \\
&= \frac{1}{2}x + \frac{1}{4}\sin 2x + C.
\end{aligned}$$

例 15 求 $\int \cos^4 x\,dx$.

解
$$\begin{aligned}
\int \cos^4 x\,dx &= \int (\cos^2 x)^2\,dx \\
&= \int \left[\frac{1}{2}(1+\cos 2x)\right]^2 dx \\
&= \frac{1}{4}\int (1 + 2\cos 2x + \cos^2 2x)\,dx \\
&= \frac{1}{4}\int \left(\frac{3}{2} + 2\cos 2x + \frac{1}{2}\cos 4x\right) dx \\
&= \frac{1}{4}\left(\frac{3}{2}x + \sin 2x + \frac{1}{8}\sin 4x\right) + C \\
&= \frac{3}{8}x + \frac{1}{4}\sin 2x + \frac{1}{32}\sin 4x + C.
\end{aligned}$$

例 16 求 $\int \sin^2 x \cos^4 x\,dx$.

解
$$\begin{aligned}
\int &\sin^2 x \cos^4 x\,dx \\
&= \frac{1}{8}\int (1-\cos 2x)(1+\cos 2x)^2\,dx \\
&= \frac{1}{8}\int (1 + \cos 2x - \cos^2 2x - \cos^3 2x)\,dx \\
&= \frac{1}{8}\int (\cos 2x - \cos^3 2x)\,dx \\
&\quad + \frac{1}{8}\int (1 - \cos^2 2x)\,dx \\
&= \frac{1}{8}\int \frac{1}{2}\sin^2 2x\,d(\sin 2x) \\
&\quad + \frac{1}{8}\int \frac{1}{2}(1-\cos 4x)\,dx \\
&= \frac{1}{48}\sin^3 2x + \frac{x}{16} - \frac{1}{64}\sin 4x + C.
\end{aligned}$$

Example 17 Compute $\int \sec^6 x \, dx$.

Solution $\int \sec^6 x \, dx = \int (\sec^2 x)^2 \sec^2 x \, dx$

$$= \int (1 + \tan^2 x)^2 \, d(\tan x)$$

$$= \int (1 + 2\tan^2 x + \tan^4 x) \, d(\tan x)$$

$$= \tan x + \frac{2}{3} \tan^3 x + \frac{1}{5} \tan^5 x + C.$$

Example 18 Compute $\int \tan^5 x \sec^3 x \, dx$.

Solution $\int \tan^5 x \sec^3 x \, dx$

$$= \int \tan^4 x \sec^2 x \sec x \tan x \, dx$$

$$= \int (\sec^2 x - 1)^2 \sec^2 x \, d(\sec x)$$

$$= \int (\sec^6 x - 2\sec^4 x + \sec^2 x) \, d(\sec x)$$

$$= \frac{1}{7} \sec^7 x - \frac{2}{5} \sec^5 x + \frac{1}{3} \sec^3 x + C.$$

Corollary 3

(1) For $\int \tan^n x \sec^{2k} x \, dx$, let $u = \tan x$;

(2) For $\int \tan^{2k-1} x \sec^n x \, dx$, let $u = \sec x$.

Example 19 Compute $\int \csc x \, dx$.

Solution $\int \csc x \, dx = \int \frac{1}{\sin x} \, dx$

$$= \int \frac{1}{2 \sin \frac{x}{2} \cos \frac{x}{2}} \, dx$$

$$= \int \frac{d\left(\frac{x}{2}\right)}{\tan \frac{x}{2} \cos^2 \frac{x}{2}}$$

$$= \int \frac{d\left(\tan \frac{x}{2}\right)}{\tan \frac{x}{2}}$$

$$= \ln \left| \tan \frac{x}{2} \right| + C$$

$$= \ln |\csc x + \cot x| + C.$$

例 17 求 $\int \sec^6 x \, dx$.

解 $\int \sec^6 x \, dx = \int (\sec^2 x)^2 \sec^2 x \, dx$

$$= \int (1 + \tan^2 x)^2 \, d(\tan x)$$

$$= \int (1 + 2\tan^2 x + \tan^4 x) \, d(\tan x)$$

$$= \tan x + \frac{2}{3} \tan^3 x + \frac{1}{5} \tan^5 x + C.$$

例 18 求 $\int \tan^5 x \sec^3 x \, dx$.

解 $\int \tan^5 x \sec^3 x \, dx$

$$= \int \tan^4 x \sec^2 x \sec x \tan x \, dx$$

$$= \int (\sec^2 x - 1)^2 \sec^2 x \, d(\sec x)$$

$$= \int (\sec^6 x - 2\sec^4 x + \sec^2 x) \, d(\sec x)$$

$$= \frac{1}{7} \sec^7 x - \frac{2}{5} \sec^5 x + \frac{1}{3} \sec^3 x + C.$$

推论 3

(1) 对于 $\int \tan^n x \sec^{2k} x \, dx$，令 $u = \tan x$；

(2) 对于 $\int \tan^{2k-1} x \sec^n x \, dx$，令 $u = \sec x$.

例 19 求 $\int \csc x \, dx$.

解 $\int \csc x \, dx = \int \frac{1}{\sin x} \, dx$

$$= \int \frac{1}{2 \sin \frac{x}{2} \cos \frac{x}{2}} \, dx$$

$$= \int \frac{d\left(\frac{x}{2}\right)}{\tan \frac{x}{2} \cos^2 \frac{x}{2}}$$

$$= \int \frac{d\left(\tan \frac{x}{2}\right)}{\tan \frac{x}{2}}$$

$$= \ln \left| \tan \frac{x}{2} \right| + C$$

$$= \ln |\csc x + \cot x| + C.$$

Example 20 Compute $\int \sec x \, dx$.

Solution
$$\int \sec x \, dx = \int \csc\left(x + \frac{\pi}{2}\right) d\left(x + \frac{\pi}{2}\right)$$
$$= \ln\left|\csc\left(x + \frac{\pi}{2}\right) - \cot\left(x + \frac{\pi}{2}\right)\right| + C$$
$$= \ln|\sec x + \tan x| + C.$$

Example 21 Compute $\int \cos 3x \cos 2x \, dx$.

Solution
$$\int \cos 3x \cos 2x \, dx$$
$$= \frac{1}{2} \int (\cos x + \cos 5x) \, dx$$
$$= \frac{1}{2} \sin x + \frac{1}{10} \sin 5x + C.$$

2. The Substitution Rule 2

Theorem 5.2 (The Substitution Rule 2) If $x = \psi(t)$ is a monotone and differentiable function, and $\psi'(t) \neq 0$, and $F(t)$ is a primitive function of $f(\psi(t))\psi'(t)$, then
$$\int f(x) \, dx = \left[\int f(\psi(t))\psi'(t) \, dt\right]_{t=\psi^{-1}(x)}$$
$$= [F(t)]_{t=\psi^{-1}(x)} + C$$
$$= F(\psi^{-1}(x)) + C,$$
where $t = \psi^{-1}(x)$ is inverse function of $x = \psi(t)$.

This is due to
$$[F(\psi^{-1}(x))]' = F'(t) \frac{dt}{dx}$$
$$= f(\psi(t))\psi'(t) \frac{1}{\frac{dx}{dt}}$$
$$= f(\psi(t)) = f(x).$$

Example 22 Compute $\int \sqrt{a^2 - x^2} \, dx \ (a > 0)$.

Solution Given $x = a\sin t \left(-\frac{\pi}{2} < t < \frac{\pi}{2}\right)$, so $\sqrt{a^2 - x^2} = \sqrt{a^2 - a^2\sin^2 t} = a\cos t$, $dx = a\cos t \, dt$, then
$$\int \sqrt{a^2 - x^2} \, dx = \int a\cos t \cdot a\cos t \, dt$$
$$= a^2 \int \cos^2 t \, dt$$
$$= a^2 \left(\frac{1}{2}t + \frac{1}{4}\sin 2t\right) + C.$$

And $t=\arcsin\dfrac{x}{a}$, $\sin 2t=2\sin t\cos t=2\dfrac{x}{a}\cdot\dfrac{\sqrt{a^2-x^2}}{a}$, so

$$\int\sqrt{a^2-x^2}\,\mathrm{d}x=a^2\left(\dfrac{1}{2}t+\dfrac{1}{4}\sin 2t\right)+C$$
$$=\dfrac{a^2}{2}\arcsin\dfrac{x}{a}+\dfrac{1}{2}x\sqrt{a^2-x^2}+C.$$

Example 23 Compute $\int\dfrac{\mathrm{d}x}{\sqrt{x^2+a^2}}$ $(a>0)$.

Solution Given $x=a\tan t\left(-\dfrac{\pi}{2}<t<\dfrac{\pi}{2}\right)$, so $\sqrt{x^2+a^2}=\sqrt{a^2+a^2\tan^2 t}=a\sqrt{1+\tan^2 t}=a\sec t$, $\mathrm{d}x=a\sec^2 t\,\mathrm{d}t$, then

$$\int\dfrac{\mathrm{d}x}{\sqrt{x^2+a^2}}=\int\dfrac{a\sec^2 t}{a\sec t}\mathrm{d}t=\int\sec t\,\mathrm{d}t$$
$$=\ln|\sec t+\tan t|+C.$$

And $\sec t=\dfrac{\sqrt{x^2+a^2}}{a}$, $\tan t=\dfrac{x}{a}$, so

$$\int\dfrac{\mathrm{d}x}{\sqrt{x^2+a^2}}=\ln|\sec t+\tan t|+C$$
$$=\ln\left(\dfrac{x}{a}+\dfrac{\sqrt{x^2+a^2}}{a}\right)+C$$
$$=\ln(x+\sqrt{x^2+a^2})+C_1,$$

where $C_1=C-\ln a$.

Example 24 Compute $\int\dfrac{\mathrm{d}x}{\sqrt{x^2-a^2}}$ $(a>0)$.

Solution If $x>a$, given $x=a\sec t\left(0<t<\dfrac{\pi}{2}\right)$, so $\sqrt{x^2-a^2}=\sqrt{a^2\sec^2 t-a^2}=a\sqrt{\sec^2 t-1}=a\tan t$, $\mathrm{d}x=a\sec t\tan t\,\mathrm{d}t$, then

$$\int\dfrac{\mathrm{d}x}{\sqrt{x^2-a^2}}=\int\dfrac{a\sec t\tan t}{a\tan t}\mathrm{d}t=\int\sec t\,\mathrm{d}t$$
$$=\ln|\sec t+\tan t|+C.$$

And $\tan t=\dfrac{\sqrt{x^2-a^2}}{a}$, $\sec t=\dfrac{x}{a}$, so

$$\int\dfrac{\mathrm{d}x}{\sqrt{x^2-a^2}}=\ln|\sec t+\tan t|+C$$
$$=\ln\left|\dfrac{x}{a}+\dfrac{\sqrt{x^2-a^2}}{a}\right|+C$$
$$=\ln(x+\sqrt{x^2-a^2})+C_1,$$

因为 $t=\arcsin\dfrac{x}{a}$, $\sin 2t=2\sin t\cos t=2\dfrac{x}{a}\cdot\dfrac{\sqrt{a^2-x^2}}{a}$, 所以

$$\int\sqrt{a^2-x^2}\,\mathrm{d}x=a^2\left(\dfrac{1}{2}t+\dfrac{1}{4}\sin 2t\right)+C$$
$$=\dfrac{a^2}{2}\arcsin\dfrac{x}{a}+\dfrac{1}{2}x\sqrt{a^2-x^2}+C.$$

例 23 求 $\int\dfrac{\mathrm{d}x}{\sqrt{x^2+a^2}}$ $(a>0)$.

解 设 $x=a\tan t\left(-\dfrac{\pi}{2}<t<\dfrac{\pi}{2}\right)$, 则 $\sqrt{x^2+a^2}=\sqrt{a^2+a^2\tan^2 t}=a\sqrt{1+\tan^2 t}=a\sec t$, $\mathrm{d}x=a\sec^2 t\,\mathrm{d}t$. 于是

$$\int\dfrac{\mathrm{d}x}{\sqrt{x^2+a^2}}=\int\dfrac{a\sec^2 t}{a\sec t}\mathrm{d}t=\int\sec t\,\mathrm{d}t.$$
$$=\ln|\sec t+\tan t|+C.$$

因为 $\sec t=\dfrac{\sqrt{x^2+a^2}}{a}$, $\tan t=\dfrac{x}{a}$, 所以

$$\int\dfrac{\mathrm{d}x}{\sqrt{x^2+a^2}}=\ln|\sec t+\tan t|+C$$
$$=\ln\left(\dfrac{x}{a}+\dfrac{\sqrt{x^2+a^2}}{a}\right)+C$$
$$=\ln(x+\sqrt{x^2+a^2})+C_1,$$

其中 $C_1=C-\ln a$.

例 24 求 $\int\dfrac{\mathrm{d}x}{\sqrt{x^2-a^2}}$ $(a>0)$.

解 若 $x>a$, 设 $x=a\sec t\left(0<t<\dfrac{\pi}{2}\right)$, 则 $\sqrt{x^2-a^2}=\sqrt{a^2\sec^2 t-a^2}=a\sqrt{\sec^2 t-1}=a\tan t$, $\mathrm{d}x=a\sec t\tan t\,\mathrm{d}t$, 于是

$$\int\dfrac{\mathrm{d}x}{\sqrt{x^2-a^2}}=\int\dfrac{a\sec t\tan t}{a\tan t}\mathrm{d}t=\int\sec t\,\mathrm{d}t$$
$$=\ln|\sec t+\tan t|+C.$$

因为 $\tan t=\dfrac{\sqrt{x^2-a^2}}{a}$, $\sec t=\dfrac{x}{a}$, 所以

$$\int\dfrac{\mathrm{d}x}{\sqrt{x^2-a^2}}=\ln|\sec t+\tan t|+C$$
$$=\ln\left|\dfrac{x}{a}+\dfrac{\sqrt{x^2-a^2}}{a}\right|+C$$
$$=\ln(x+\sqrt{x^2-a^2})+C_1,$$

where $C_1 = C - \ln a$.

If $x < a$, given $x = -u$, so $u > a$, and
$$\int \frac{dx}{\sqrt{x^2-a^2}} = -\int \frac{du}{\sqrt{u^2-a^2}}$$
$$= -\ln(u+\sqrt{u^2-a^2}) + C$$
$$= -\ln(-x+\sqrt{x^2-a^2}) + C$$
$$= \ln \frac{-x-\sqrt{x^2-a^2}}{a^2} + C$$
$$= \ln(-x-\sqrt{x^2-a^2}) + C_1,$$

where $C_1 = C - 2\ln a$.

In summary, we have
$$\int \frac{dx}{\sqrt{x^2-a^2}} = \ln|x+\sqrt{x^2-a^2}| + C.$$

Some more Basic formulas of Integrals

(16) $\int \tan x \, dx = -\ln|\cos x| + C;$

(17) $\int \cot x \, dx = \ln|\sin x| + C;$

(18) $\int \sec x \, dx = \ln|\sec x + \tan x| + C;$

(19) $\int \csc x \, dx = \ln|\csc x - \cot x| + C;$

(20) $\int \frac{1}{a^2+x^2} dx = \frac{1}{a}\arctan \frac{x}{a} + C;$

(21) $\int \frac{1}{x^2-a^2} dx = \frac{1}{2a}\ln\left|\frac{x-a}{x+a}\right| + C;$

(22) $\int \frac{1}{\sqrt{a^2-x^2}} dx = \arcsin \frac{x}{a} + C;$

(23) $\int \frac{1}{\sqrt{x^2+a^2}} dx = \ln(x+\sqrt{x^2+a^2}) + C;$

(24) $\int \frac{1}{\sqrt{x^2-a^2}} dx = \ln|x+\sqrt{x^2-a^2}| + C.$

Example 25 Compute $\int \frac{dx}{\sqrt{4x^2+9}}$.

Solution $\int \frac{dx}{\sqrt{4x^2+9}} = \frac{1}{2}\int \frac{d(2x)}{\sqrt{(2x)^2+3^2}}$
$$= \frac{1}{2}\ln(2x+\sqrt{4x^2+9}) + C.$$

Example 26 Compute $\int \frac{dx}{\sqrt{1+x-x^2}}$.

Solution $\int \dfrac{dx}{\sqrt{1+x-x^2}} = \int \dfrac{d\left(x-\dfrac{1}{2}\right)}{\sqrt{\left(\dfrac{\sqrt{5}}{2}\right)^2 - \left(x-\dfrac{1}{2}\right)^2}}$

$= \arcsin \dfrac{2x-1}{\sqrt{5}} + C.$

5.3 Integration by Parts
5.3 分部积分法

If the derivatives of the functions $u=u(x)$ and $v=v(x)$ are continuous, and the derivative formula of the product of two functions is
$$(uv)' = u'v + uv',$$
or
$$uv' = (uv)' - u'v,$$
evaluating indefinite integral in this expression, we know that
$$\int uv' dx = uv - \int u'v dx$$
or
$$\int u dv = uv - \int v du.$$

This is the **formula of integration by parts**.

Inferential expression is
$$\int uv' dx = \int u dv = uv - \int v du$$
$$= uv - \int u'v dx = \cdots.$$

Principle: v is easy to compute, and computing $\int v du$ is easier than computing $\int u dv$.

Example 1 Compute $\int x\cos x dx.$

Solution $\int x\cos x dx = \int x d(\sin x)$
$= x\sin x - \int \sin x dx$
$= x\sin x + \cos x + C.$

Example 2 Compute $\int xe^x dx.$

Solution $\int xe^x dx = \int x d(e^x) = xe^x - \int e^x dx$
$= xe^x - e^x + C.$

如果函数 $u=u(x)$ 及 $v=v(x)$ 具有连续导数,那么两个函数乘积的导数公式为
$$(uv)' = u'v + uv'$$
或
$$uv' = (uv)' - u'v.$$
对这个等式两边求不定积分,得
$$\int uv' dx = uv - \int u'v dx$$

或 $\quad \int u dv = uv - \int v du.$

这个公式就是**分部积分法公式**.

分部积分法可表示为
$$\int uv' dx = \int u dv = uv - \int v du$$
$$= uv - \int u'v dx = \cdots.$$

原则:v 容易求出,且 $\int v du$ 比 $\int u dv$ 容易计算.

例1 求 $\int x\cos x dx.$

解 $\int x\cos x dx = \int x d(\sin x)$
$= x\sin x - \int \sin x dx$
$= x\sin x + \cos x + C.$

例2 求 $\int xe^x dx.$

解 $\int xe^x dx = \int x d(e^x) = xe^x - \int e^x dx$
$= xe^x - e^x + C.$

Example 3 Compute $\int x^2 e^x dx$.

Solution
$$\int x^2 e^x dx = \int x^2 d(e^x)$$
$$= x^2 e^x - \int e^x d(x^2)$$
$$= x^2 e^x - 2\int x e^x dx$$
$$= x^2 e^x - 2\int x d(e^x)$$
$$= x^2 e^x - 2xe^x + 2\int e^x dx$$
$$= x^2 e^x - 2xe^x + 2e^x + C$$
$$= e^x(x^2 - 2x + 2) + C.$$

Example 4 Compute $\int x \ln x dx$.

Solution
$$\int x \ln x dx = \frac{1}{2}\int \ln x d(x^2)$$
$$= \frac{1}{2}x^2 \ln x - \frac{1}{2}\int x^2 \cdot \frac{1}{x} dx$$
$$= \frac{1}{2}x^2 \ln x - \frac{1}{2}\int x dx$$
$$= \frac{1}{2}x^2 \ln x - \frac{1}{4}x^2 + C.$$

Example 5 Compute $\int \arccos x dx$.

Solution
$$\int \arccos x dx = x\arccos x - \int x d(\arccos x)$$
$$= x\arccos x + \int x \frac{1}{\sqrt{1-x^2}} dx$$
$$= x\arccos x - \frac{1}{2}\int (1-x^2)^{-\frac{1}{2}} d(1-x^2)$$
$$= x\arccos x - \sqrt{1-x^2} + C.$$

Example 6 Compute $\int x\arctan x dx$.

Solution
$$\int x\arctan x dx = \frac{1}{2}\int \arctan x d(x^2)$$
$$= \frac{1}{2}x^2 \arctan x - \frac{1}{2}\int x^2 \cdot \frac{1}{1+x^2} dx$$
$$= \frac{1}{2}x^2 \arctan x - \frac{1}{2}\int \left(1 - \frac{1}{1+x^2}\right) dx$$
$$= \frac{1}{2}x^2 \arctan x - \frac{1}{2}x + \frac{1}{2}\arctan x + C.$$

Example 7 Compute $\int e^x \sin x dx$.

Solution As

$$\int e^x \sin x \, dx = \int \sin x \, d(e^x)$$
$$= e^x \sin x - \int e^x \, d(\sin x)$$
$$= e^x \sin x - \int e^x \cos x \, dx$$
$$= e^x \sin x - \int \cos x \, d(e^x)$$
$$= e^x \sin x - e^x \cos x + \int e^x \, d(\cos x)$$
$$= e^x \sin x - e^x \cos x - \int e^x \sin x \, dx,$$

so

$$\int e^x \sin x \, dx = \frac{1}{2} e^x (\sin x - \cos x) + C.$$

Example 8 Compute $\int \sec^3 x \, dx$.

Solution As

$$\int \sec^3 x \, dx = \int \sec x \cdot \sec^2 x \, dx$$
$$= \int \sec x \, d(\tan x)$$
$$= \sec x \tan x - \int \sec x \tan^2 x \, dx$$
$$= \sec x \tan x - \int \sec x (\sec^2 x - 1) \, dx$$
$$= \sec x \tan x - \int \sec^3 x \, dx + \int \sec x \, dx$$
$$= \sec x \tan x + \ln|\sec x + \tan x|$$
$$\quad - \int \sec^3 x \, dx,$$

so

$$\int \sec^3 x \, dx = \frac{1}{2} (\sec x \tan x + \ln|\sec x + \tan x|) + C.$$

Example 9 Compute $\int e^{\sqrt{x}} \, dx$.

Solution Given $\sqrt{x} = t$, $x = t^2$, so $dx = 2t \, dt$, then

$$\int e^{\sqrt{x}} \, dx = 2 \int t e^t \, dt$$
$$= 2 e^t (t - 1) + C$$
$$= 2 e^{\sqrt{x}} (\sqrt{x} - 1) + C.$$

5.4 The Indefinite Integral of the Rational Function
5.4 有理函数的不定积分

1. The Indefinite Integral of the Rational Function

The **rational function** is a function who is expressed by the quotient of two polynomials, or which with the following form:
$$\frac{P(x)}{Q(x)} = \frac{a_0 x^n + a_1 x^{n-1} + \cdots + a_{n-1} x + a_n}{b_0 x^m + b_1 x^{m-1} + \cdots + b_{m-1} x + b_m},$$
where m and n are non-negative integer, $a_0, a_1, a_2, \cdots, a_n$ and $b_0, b_1, b_2, \cdots, b_m$ are real numbers, and $a_0 \neq 0$, $b_0 \neq 0$. If $n < m$, this rational function is called the **proper fraction**, else called the **improper fraction**.

Improper fraction could expressed by a polynomial add a proper fraction. For example,
$$\frac{x^3 + x + 1}{x^2 + 1} = \frac{x(x^2 + 1) + 1}{x^2 + 1}$$
$$= x + \frac{1}{x^2 + 1}.$$

When compute an indefinite integral of proper fraction, if the denominator could be factored, factor it first, then compute the indefinite integral of partial fraction.

Example 1 Compute $\int \frac{x+1}{x^2 - 5x + 6} dx$.

Solution Let
$$\frac{x+1}{(x-2)(x-3)} = \frac{A}{x-3} + \frac{B}{x-2}$$
$$= \frac{(A+B)x + (-2A - 3B)}{(x-2)(x-3)},$$
then
$$A + B = 1, \quad -2A - 3B = 1,$$
$$A = 4, \quad B = -3,$$
so
$$\int \frac{x+1}{x^2 - 5x + 6} dx = \int \frac{x+3}{(x-2)(x-3)} dx$$
$$= \int \left(\frac{4}{x-3} - \frac{3}{x-2} \right) dx$$
$$= \int \frac{4}{x-3} dx - \int \frac{3}{x-2} dx$$
$$= 4\ln|x-3| - 3\ln|x-2| + C.$$

1. 有理函数的不定积分

有理函数是指由两个多项式的商所表示的函数,即具有如下形式的函数:
$$\frac{P(x)}{Q(x)} = \frac{a_0 x^n + a_1 x^{n-1} + \cdots + a_{n-1} x + a_n}{b_0 x^m + b_1 x^{m-1} + \cdots + b_{m-1} x + b_m},$$
其中 m 和 n 都是非负整数,$a_0, a_1, a_2, \cdots, a_n$ 及 $b_0, b_1, b_2, \cdots, b_m$ 都是实数,并且 $a_0 \neq 0$,$b_0 \neq 0$. 当 $n < m$ 时,称这一有理函数为**真分式**;而当 $n \geq m$ 时,称这一有理函数为**假分式**.

假分式总可以化成一个多项式与一个真分式之和的形式. 例如:
$$\frac{x^3 + x + 1}{x^2 + 1} = \frac{x(x^2 + 1) + 1}{x^2 + 1}$$
$$= x + \frac{1}{x^2 + 1}.$$

求真分式的不定积分时,如果分母可因式分解,则先因式分解,然后化成部分分式再积分.

例1 求 $\int \frac{x+1}{x^2 - 5x + 6} dx$.

解 令
$$\frac{x+1}{(x-2)(x-3)} = \frac{A}{x-3} + \frac{B}{x-2}$$
$$= \frac{(A+B)x + (-2A - 3B)}{(x-2)(x-3)},$$
则有
$$A + B = 1, \quad -2A - 3B = 1,$$
$$A = 4, \quad B = -3.$$
于是
$$\int \frac{x+1}{x^2 - 5x + 6} dx = \int \frac{x+3}{(x-2)(x-3)} dx$$
$$= \int \left(\frac{4}{x-3} - \frac{3}{x-2} \right) dx$$
$$= \int \frac{4}{x-3} dx - \int \frac{3}{x-2} dx$$
$$= 4\ln|x-3| - 3\ln|x-2| + C.$$

5.4 The Indefinite Integral of the Rational Function
5.4 有理函数的不定积分

Example 2 Compute $\int \dfrac{x-2}{x^2+2x+3}\mathrm{d}x$.

Solution $\int \dfrac{x-2}{x^2+2x+3}\mathrm{d}x$

$= \int \left(\dfrac{1}{2} \cdot \dfrac{2x+2}{x^2+2x+3} - 3 \cdot \dfrac{1}{x^2+2x+3} \right) \mathrm{d}x$

$= \dfrac{1}{2} \int \dfrac{2x+2}{x^2+2x+3} \mathrm{d}x - 3 \int \dfrac{1}{x^2+2x+3} \mathrm{d}x$

$= \dfrac{1}{2} \int \dfrac{\mathrm{d}(x^2+2x+3)}{x^2+2x+3} - 3 \int \dfrac{\mathrm{d}(x+1)}{(x+1)^2+(\sqrt{2})^2}$

$= \dfrac{1}{2} \ln(x^2+2x+3) - \dfrac{3}{\sqrt{2}} \arctan \dfrac{x+1}{\sqrt{2}} + C.$

Guide $\dfrac{x-2}{x^2+2x+3} = \dfrac{\dfrac{1}{2}(2x+2)-3}{x^2+2x+3}$

$= \dfrac{1}{2} \cdot \dfrac{2x+2}{x^2+2x+3} - 3 \cdot \dfrac{1}{x^2+2x+3}.$

Example 3 Compute $\int \dfrac{1}{x(x-1)^2}\mathrm{d}x$.

Solution $\int \dfrac{1}{x(x-1)^2} \mathrm{d}x$

$= \int \left[\dfrac{1}{x} - \dfrac{1}{x-1} + \dfrac{1}{(x-1)^2} \right] \mathrm{d}x$

$= \int \dfrac{1}{x} \mathrm{d}x - \int \dfrac{1}{x-1} \mathrm{d}x + \int \dfrac{1}{(x-1)^2} \mathrm{d}x$

$= \ln|x| - \ln|x-1| - \dfrac{1}{x-1} + C.$

Guide $\dfrac{1}{x(x-1)^2} = \dfrac{1-x+x}{x(x-1)^2}$

$= -\dfrac{1}{x(x-1)} + \dfrac{1}{(x-1)^2}$

$= -\dfrac{1-x+x}{x(x-1)} + \dfrac{1}{(x-1)^2}$

$= \dfrac{1}{x} - \dfrac{1}{x-1} + \dfrac{1}{(x-1)^2}.$

例 2 求 $\int \dfrac{x-2}{x^2+2x+3}\mathrm{d}x$.

解 $\int \dfrac{x-2}{x^2+2x+3}\mathrm{d}x$

$= \int \left(\dfrac{1}{2} \cdot \dfrac{2x+2}{x^2+2x+3} - 3 \cdot \dfrac{1}{x^2+2x+3} \right) \mathrm{d}x$

$= \dfrac{1}{2} \int \dfrac{2x+2}{x^2+2x+3} \mathrm{d}x - 3 \int \dfrac{1}{x^2+2x+3} \mathrm{d}x$

$= \dfrac{1}{2} \int \dfrac{\mathrm{d}(x^2+2x+3)}{x^2+2x+3} - 3 \int \dfrac{\mathrm{d}(x+1)}{(x+1)^2+(\sqrt{2})^2}$

$= \dfrac{1}{2} \ln(x^2+2x+3) - \dfrac{3}{\sqrt{2}} \arctan \dfrac{x+1}{\sqrt{2}} + C.$

提示 $\dfrac{x-2}{x^2+2x+3} = \dfrac{\dfrac{1}{2}(2x+2)-3}{x^2+2x+3}$

$= \dfrac{1}{2} \cdot \dfrac{2x+2}{x^2+2x+3} - 3 \cdot \dfrac{1}{x^2+2x+3}.$

例 3 求 $\int \dfrac{1}{x(x-1)^2}\mathrm{d}x$.

解 $\int \dfrac{1}{x(x-1)^2} \mathrm{d}x$

$= \int \left[\dfrac{1}{x} - \dfrac{1}{x-1} + \dfrac{1}{(x-1)^2} \right] \mathrm{d}x$

$= \int \dfrac{1}{x} \mathrm{d}x - \int \dfrac{1}{x-1} \mathrm{d}x + \int \dfrac{1}{(x-1)^2} \mathrm{d}x$

$= \ln|x| - \ln|x-1| - \dfrac{1}{x-1} + C.$

提示 $\dfrac{1}{x(x-1)^2} = \dfrac{1-x+x}{x(x-1)^2}$

$= -\dfrac{1}{x(x-1)} + \dfrac{1}{(x-1)^2}$

$= -\dfrac{1-x+x}{x(x-1)} + \dfrac{1}{(x-1)^2}$

$= \dfrac{1}{x} - \dfrac{1}{x-1} + \dfrac{1}{(x-1)^2}.$

2. The Indefinite Integral of the Rational Function with Trigonometric Function

A rational function with trigonometric function is a function constituted by trigonometric function and constant, with limited arithmetic. Its numerator and denominator are include arithmetic on trigonometric function. As trigonometric functions all could be expressed by the rational function with $\sin x$ and $\cos x$, so a rational function with trigonometric function is a rational function with $\sin x$ and $\cos x$, too.

For an indefinite integral of rational function with trigonometric function, expressed $\sin x$ and $\cos x$ by $\tan\frac{x}{2}$, given $u = \tan\frac{x}{2}$:

$$\sin x = 2\sin\frac{x}{2}\cos\frac{x}{2} = \frac{2\tan\frac{x}{2}}{\sec^2\frac{x}{2}}$$

$$= \frac{2\tan\frac{x}{2}}{1+\tan^2\frac{x}{2}} = \frac{2u}{1+u^2},$$

$$\cos x = \cos^2\frac{x}{2} - \sin^2\frac{x}{2} = \frac{1-\tan^2\frac{x}{2}}{\sec^2\frac{x}{2}} = \frac{1-u^2}{1+u^2},$$

then it becomes an integral of rational function.

Example 4 Compute $\int \frac{1+\sin x}{\sin x(1+\cos x)}dx$.

Solution Given $u = \tan\frac{x}{2}$, then $\sin x = \frac{2u}{1+u^2}$, $\cos x = \frac{1-u^2}{1+u^2}$, $x = 2\arctan u$, $dx = \frac{2}{1+u^2}du$, so

$$\int \frac{1+\sin x}{\sin x(1+\cos x)}dx$$

$$= \int \frac{\left(1+\frac{2u}{1+u^2}\right)}{\frac{2u}{1+u^2}\left(1+\frac{1-u^2}{1+u^2}\right)} \cdot \frac{2}{1+u^2}du$$

$$= \frac{1}{2}\int \left(u+2+\frac{1}{u}\right)du$$

2. 三角函数有理式的不定积分

三角函数有理式是指由三角函数和常数经过有限次四则运算所构成的有理函数,其特点是分子、分母都包含三角函数的和、差和乘积运算.由于各种三角函数都可以用 $\sin x$ 及 $\cos x$ 的有理式表示,故三角函数有理式也就是 $\sin x, \cos x$ 的有理式.

对于三角函数有理式的不定积分,把 $\sin x, \cos x$ 表示成 $\tan\frac{x}{2}$ 的函数,然后作替换 $u = \tan\frac{x}{2}$:

$$\sin x = 2\sin\frac{x}{2}\cos\frac{x}{2} = \frac{2\tan\frac{x}{2}}{\sec^2\frac{x}{2}}$$

$$= \frac{2\tan\frac{x}{2}}{1+\tan^2\frac{x}{2}} = \frac{2u}{1+u^2},$$

$$\cos x = \cos^2\frac{x}{2} - \sin^2\frac{x}{2} = \frac{1-\tan^2\frac{x}{2}}{\sec^2\frac{x}{2}} = \frac{1-u^2}{1+u^2},$$

则原不定积分变成了有理函数的不定积分.

例 4 求 $\int \frac{1+\sin x}{\sin x(1+\cos x)}dx$.

解 令 $u = \tan\frac{x}{2}$,则 $\sin x = \frac{2u}{1+u^2}$, $\cos x = \frac{1-u^2}{1+u^2}$, $x = 2\arctan u$, $dx = \frac{2}{1+u^2}du$.于是

$$\int \frac{1+\sin x}{\sin x(1+\cos x)}dx$$

$$= \int \frac{\left(1+\frac{2u}{1+u^2}\right)}{\frac{2u}{1+u^2}\left(1+\frac{1-u^2}{1+u^2}\right)} \cdot \frac{2}{1+u^2}du$$

$$= \frac{1}{2}\int \left(u+2+\frac{1}{u}\right)du$$

$$= \frac{1}{2}\left(\frac{u^2}{2}+2u+\ln|u|\right)+C$$
$$= \frac{1}{4}\tan^2\frac{x}{2}+\tan\frac{x}{2}$$
$$+\frac{1}{2}\ln\left|\tan\frac{x}{2}\right|+C.$$

Remark Not all indefinite integral of rational function with trigonometric function are need to transform to indefinite integral of rational function. For example,
$$\int \frac{\cos x}{1+\sin x}dx = \int \frac{1}{1+\sin x}d(1+\sin x)$$
$$=\ln(1+\sin x)+C.$$

3. The Indefinite Integral of the Simple Irrational Function

In most cases, an indefinite integral of irrational function need the Substitution Rule 2 to eliminate the sign of evolution.

Example 5 Compute $\int \frac{\sqrt{x-1}}{x}dx$.

Solution Given $\sqrt{x-1}=u$, or $x=u^2+1$, then
$$\int \frac{\sqrt{x-1}}{x}dx = \int \frac{u}{u^2+1}\cdot 2u\,du$$
$$= 2\int \frac{u^2}{u^2+1}du$$
$$= 2\int \left(1-\frac{1}{1+u^2}\right)du$$
$$= 2(u-\arctan u)+C$$
$$= 2(\sqrt{x-1}-\arctan\sqrt{x-1})+C.$$

Example 6 Compute $\int \frac{dx}{1+\sqrt[3]{x+2}}$.

Solution Given $\sqrt[3]{x+2}=u$, or $x=u^3-2$, then
$$\int \frac{dx}{1+\sqrt[3]{x+2}} = \int \frac{1}{1+u}\cdot 3u^2\,du$$
$$= 3\int \frac{u^2-1+1}{1+u}du$$
$$= 3\int \left(u-1+\frac{1}{1+u}\right)du$$
$$= 3\left(\frac{u^2}{2}-u+\ln|1+u|\right)+C$$
$$= \frac{3}{2}\sqrt[3]{(x+2)^2}-3\sqrt[3]{x+2}$$
$$+\ln|1+\sqrt[3]{x+2}|+C.$$

$$= \frac{1}{2}\left(\frac{u^2}{2}+2u+\ln|u|\right)+C$$
$$= \frac{1}{4}\tan^2\frac{x}{2}+\tan\frac{x}{2}$$
$$+\frac{1}{2}\ln\left|\tan\frac{x}{2}\right|+C.$$

注 并非所有的三角函数有理式的不定积分都要通过变换化为有理函数的不定积分. 例如:
$$\int \frac{\cos x}{1+\sin x}dx = \int \frac{1}{1+\sin x}d(1+\sin x)$$
$$=\ln(1+\sin x)+C.$$

3. 简单无理函数的不定积分

大多数情况下,无理函数的不定积分一般要采用第二换元法把根号消去.

例 5 求 $\int \frac{\sqrt{x-1}}{x}dx$.

解 设 $\sqrt{x-1}=u$,即 $x=u^2+1$,则
$$\int \frac{\sqrt{x-1}}{x}dx = \int \frac{u}{u^2+1}\cdot 2u\,du$$
$$= 2\int \frac{u^2}{u^2+1}du$$
$$= 2\int \left(1-\frac{1}{1+u^2}\right)du$$
$$= 2(u-\arctan u)+C$$
$$= 2(\sqrt{x-1}-\arctan\sqrt{x-1})+C.$$

例 6 求 $\int \frac{dx}{1+\sqrt[3]{x+2}}$.

解 设 $\sqrt[3]{x+2}=u$,即 $x=u^3-2$,则
$$\int \frac{dx}{1+\sqrt[3]{x+2}} = \int \frac{1}{1+u}\cdot 3u^2\,du$$
$$= 3\int \frac{u^2-1+1}{1+u}du$$
$$= 3\int \left(u-1+\frac{1}{1+u}\right)du$$
$$= 3\left(\frac{u^2}{2}-u+\ln|1+u|\right)+C$$
$$= \frac{3}{2}\sqrt[3]{(x+2)^2}-3\sqrt[3]{x+2}$$
$$+\ln|1+\sqrt[3]{x+2}|+C.$$

Example 7 Compute $\int \dfrac{dx}{(1+\sqrt[3]{x})\sqrt{x}}$.

Solution Given $x=t^6$, then $dx=6t^5 dt$, so

$$\int \dfrac{dx}{(1+\sqrt[3]{x})\sqrt{x}} = \int \dfrac{6t^5}{(1+t^2)t^3} dt = 6\int \dfrac{t^2}{1+t^2} dt$$

$$= 6\int \left(1-\dfrac{1}{1+t^2}\right)dt = 6(t-\arctan t)+C$$

$$= 6(\sqrt[6]{x}-\arctan \sqrt[6]{x})+C.$$

Example 8 Compute $\int \dfrac{1}{x}\sqrt{\dfrac{1+x}{x}}dx$.

Solution Given $\sqrt{\dfrac{1+x}{x}}=t$, or $x=\dfrac{1}{t^2-1}$, then

$$\int \dfrac{1}{x}\sqrt{\dfrac{1+x}{x}}dx = \int (t^2-1)t \cdot \dfrac{-2t}{(t^2-1)^2} dt$$

$$= -2\int \dfrac{t^2}{t^2-1} dt = -2\int \left(1+\dfrac{1}{t^2-1}\right)dt$$

$$= -2t-\ln\left|\dfrac{t-1}{t+1}\right|+C$$

$$= -2\sqrt{\dfrac{1+x}{x}}-\ln\dfrac{\sqrt{1+x}-\sqrt{x}}{\sqrt{1+x}+\sqrt{x}}+C.$$

Exercises 5

1. Compute the following indefinite integrals:

(1) $\int \dfrac{dx}{x^2}$;

(2) $\int 2x^4\sqrt[3]{x}\,dx$;

(3) $\int (x^3+2)^2 dx$;

(4) $\int \left(e^x+\dfrac{5}{x}\right)dx$;

(5) $\int \left(\dfrac{4}{1+x^2}+\dfrac{3}{\sqrt{1-x^2}}\right)dx$;

(6) $\int \dfrac{dt}{1+\cos 2t}$.

2. Applying the Substitution Rule 1, compute the following indefinite integrals:

(1) $\int (2-5x)^3 dx$;

(2) $\int \dfrac{\mathrm{d}x}{\sqrt{3-2x}}$;

(3) $\int \dfrac{\mathrm{d}x}{4-x}$;

(4) $\int x^2 \mathrm{e}^{3x^3} \mathrm{d}x$;

(5) $\int \dfrac{x^6}{x^7+2} \mathrm{d}x$.

3. Applying the Substitution Rule 2, find the following indefinite integrals:

(1) $\int \dfrac{\mathrm{d}x}{1+\sqrt{x+2}}$;

(2) $\int \dfrac{\mathrm{d}x}{2+\cos x}$;

(3) $\int \dfrac{\mathrm{d}x}{\sqrt{1-x}+\sqrt[4]{1-x}}$;

(4) $\int \dfrac{\mathrm{d}x}{x^3(x^2+1)}$;

(5) $\int \dfrac{\mathrm{d}x}{\sqrt[3]{(1-x)^2(1+x)^4}}$.

4. Applying partial integral method, find the following indefinite integrals:

(1) $\int x\cos 2x \, \mathrm{d}x$;

(2) $\int (x-1)\cos 5x \, \mathrm{d}x$;

(3) $\int x\mathrm{e}^{-x} \mathrm{d}x$;

(4) $\int (2x-1)\ln x \, \mathrm{d}x$;

(5) $\int x^2 \ln 3x \, \mathrm{d}x$.

5. Find the following indefinite integrals of the rational functions:

(1) $\int \dfrac{x^2}{x+1} \mathrm{d}x$;

(2) $\int \dfrac{x^3}{x+2} \mathrm{d}x$;

(3) $\int \dfrac{2x-3}{x^2-3x+5} \mathrm{d}x$;

(4) $\int \dfrac{x-2}{x^2-6x+10} \mathrm{d}x$;

(5) $\int \dfrac{\mathrm{d}x}{x(x^2+3)}$.

3. 应用第二换元法求下列不定积分:

(1) $\int \dfrac{\mathrm{d}x}{1+\sqrt{x+2}}$;

(2) $\int \dfrac{\mathrm{d}x}{2+\cos x}$;

(3) $\int \dfrac{\mathrm{d}x}{\sqrt{1-x}+\sqrt[4]{1-x}}$;

(4) $\int \dfrac{\mathrm{d}x}{x^3(x^2+1)}$;

(5) $\int \dfrac{\mathrm{d}x}{\sqrt[3]{(1-x)^2(1+x)^4}}$.

4. 应用分部积分法求下列不定积分:

(1) $\int x\cos 2x \, \mathrm{d}x$;

(2) $\int (x-1)\cos 5x \, \mathrm{d}x$;

(3) $\int x\mathrm{e}^{-x} \mathrm{d}x$;

(4) $\int (2x-1)\ln x \, \mathrm{d}x$;

(5) $\int x^2 \ln 3x \, \mathrm{d}x$.

5. 求下列有理函数的不定积分:

(1) $\int \dfrac{x^2}{x+1} \mathrm{d}x$;

(2) $\int \dfrac{x^3}{x+2} \mathrm{d}x$;

(3) $\int \dfrac{2x-3}{x^2-3x+5} \mathrm{d}x$;

(4) $\int \dfrac{x-2}{x^2-6x+10} \mathrm{d}x$;

(5) $\int \dfrac{\mathrm{d}x}{x(x^2+3)}$.

Chapter 6　The Definite Integral
第6章　定积分

6.1　The Concept and the Properties of the Definite Integral
6.1　定积分的概念与性质

1. Examples of the Definite Integral

1) Area of the Trapezoid with Curve Side

Suppose that the function $y=f(x)$ is continuous and nonnegative on the interval $[a,b]$. The graph enclosed by the line $x=a$, $x=b$, $y=0$ and the curve $y=f(x)$ is shown in Figure 6.1(a). The graph looked like this is called a **trapezoid with curve**, and the curve is called the **curve side**.

Now, we compute the area of the above-mentioned trapezoid with curve (Figure 6.1(b)):

(1) Using division point
$$a=x_0<x_1<x_2<\cdots<x_{n-1}<x_n=b,$$
we divide the interval $[a,b]$ into n sub-intervals:
$$[x_0,x_1],\ [x_1,x_2],\ \cdots,\ [x_{n-1},x_n].$$
Sign
$$\Delta x_i=x_i-x_{i-1}\quad (i=1,2,\cdots,n).$$

1. 定积分问题举例

1) 曲边梯形的面积

设函数 $y=f(x)$ 在区间 $[a,b]$ 上非负、连续. 由直线 $x=a, x=b, y=0$ 及曲线 $y=f(x)$ 所围成的图形如图 6.1(a) 所示, 像这样的图形称为**曲边梯形**, 其中曲线弧称为**曲边**.

现在我们来求上述曲边梯形的面积(图 6.1(b)):

(1) 用分点
$$a=x_0<x_1<x_2<\cdots<x_{n-1}<x_n=b$$
把区间 $[a,b]$ 分成 n 个小区间:
$$[x_0,x_1],\ [x_1,x_2],\ \cdots,\ [x_{n-1},x_n].$$
记
$$\Delta x_i=x_i-x_{i-1}\quad (i=1,2,\cdots,n).$$

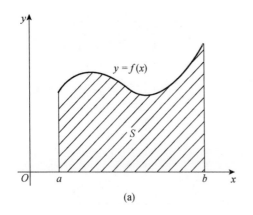
(a)

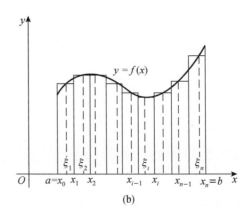
(b)

Figure 6.1

图 6.1

(2) Take any $\xi_i \in [x_{i-1}, x_i]$, then the area of the trapezoid with curve based $[x_{i-1}, x_i]$ is approximate to
$$f(\xi_i)\Delta x_i \quad (i=1,2,\cdots,n).$$

(3) The area S of the trapezoid with curve side is approximate to
$$S \approx \sum_{i=1}^{n} f(\xi_i)\Delta x_i.$$

(4) Sign $\lambda = \max\{\Delta x_1, \Delta x_2, \cdots, \Delta x_n\}$, so the exact area of the trapezoid with curve is
$$S = \lim_{\lambda \to 0} \sum_{i=1}^{n} f(\xi_i)\Delta x_i.$$

2) Distance of Variable Speed Liner Motion

Let the object move in a straight line, it is known that the speed $v = v(t)$ is a continuous function of t in time interval $[T_1, T_2]$, and $v(t) \geqslant 0$, then calculate the distance s of an object during this time.

(1) Using the point
$$T_1 = t_0 < t_1 < t_2 < \cdots < t_{n-1} < t_n = T_2,$$
we divide the time interval $[T_1, T_2]$ into n small time intervals:
$$[t_0, t_1], \quad [t_1, t_2], \quad \cdots, \quad [t_{n-1}, t_n].$$
Sign $\Delta t_i = t_i - t_{i-1} (i=1,2,\cdots,n)$.

(2) Take any $\tau_i \in [t_{i-1}, t_i]$, then the distance of an object during the time interval $[t_{i-1}, t_i]$ is approximated to
$$v(\tau_i)\Delta t_i \quad (i=1,2,\cdots,n).$$

(3) The distance s is approximated to
$$s \approx \sum_{i=1}^{n} v(\tau_i)\Delta t_i.$$

(4) Sign $\lambda = \max\{\Delta t_1, \Delta t_2, \cdots, \Delta t_n\}$, then the exact distance is
$$s = \lim_{\lambda \to 0} \sum_{i=1}^{n} v(\tau_i)\Delta t_i.$$

2. The Definition of the Definite Integral

Aside the specific meaning of the above-mentioned problems, grasp the quantitative relation of the common essence and characteristics, we can define the definite integral as follow:

（2）任取 $\xi_i \in [x_{i-1}, x_i]$，以 $[x_{i-1}, x_i]$ 为底的小曲边梯形的面积可近似为
$$f(\xi_i)\Delta x_i \quad (i=1,2,\cdots,n);$$

（3）所求曲边梯形面积 S 的近似值为
$$S \approx \sum_{i=1}^{n} f(\xi_i)\Delta x_i.$$

（4）记 $\lambda = \max\{\Delta x_1, \Delta x_2, \cdots, \Delta x_n\}$，所以曲边梯形面积的精确值为
$$S = \lim_{\lambda \to 0} \sum_{i=1}^{n} f(\xi_i)\Delta x_i.$$

2) 变速直线运动的路程

设物体做直线运动，已知速度 $v = v(t)$ 是时间区间 $[T_1, T_2]$ 上 t 的连续函数，且 $v(t) \geqslant 0$，计算在这段时间内物体所经过的路程 s。

（1）用分点
$$T_1 = t_0 < t_1 < t_2 < \cdots < t_{n-1} < t_n = T_2$$
把时间区间 $[T_1, T_2]$ 分成 n 个小时间段：
$$[t_0, t_1], \quad [t_1, t_2], \quad \cdots, \quad [t_{n-1}, t_n].$$
记 $\Delta t_i = t_i - t_{i-1} (i=1,2,\cdots,n)$。

（2）任取 $\tau_i \in [t_{i-1}, t_i]$，在时间段 $[t_{i-1}, t_i]$ 内物体所经过的路程可近似为
$$v(\tau_i)\Delta t_i \quad (i=1,2,\cdots,n).$$

（3）所求路程 s 的近似值为
$$s \approx \sum_{i=1}^{n} v(\tau_i)\Delta t_i.$$

（4）记 $\lambda = \max\{\Delta t_1, \Delta t_2, \cdots, \Delta t_n\}$，则所求路程的精确值为
$$s = \lim_{\lambda \to 0} \sum_{i=1}^{n} v(\tau_i)\Delta t_i.$$

2. 定积分的定义

抛开上述问题的具体意义，抓住它们在数量关系上共同的本质与特性加以概括，就抽象出下述定积分的定义：

Chapter 6 The Definite Integral
第 6 章 定积分

Definition 6.1 Suppose that the function $f(x)$ is bounded in $[a,b]$. Insert $n-1$ points,
$$a=x_0<x_1<x_2<\cdots<x_{n-1}<x_n=b,$$
into $[a,b]$ and divide it into n sub-intervals:
$$[x_0,x_1], [x_1,x_2], \cdots, [x_{n-1},x_n].$$
The length of small intervals is
$$\Delta x_1=x_1-x_0, \Delta x_2=x_2-x_1,$$
$$\cdots, \Delta x_n=x_n-x_{n-1}.$$
Take a point ξ_i in every sub-intervals $[x_{i-1},x_i]$, compute the product of the function value $f(\xi_i)$ and the length of sub-interval Δx_i,
$$f(\xi_i)\Delta x_i \quad (i=1,2,\cdots,n),$$
and compute
$$A=\sum_{i=1}^{n}f(\xi_i)\Delta x_i.$$
Sign $\lambda=\max\{\Delta x_1,\Delta x_2,\cdots,\Delta x_n\}$. No matter how to divide $[a,b]$ and how to choose the point ξ_i in sub-interval $[x_{i-1},x_i]$, as long as $\lambda \to 0$, the sum A tend to an certain limit I, then we called this limit I as the **definite integral** of the function $f(x)$ on the interval $[a,b]$, signed as $\int_a^b f(x)\mathrm{d}x$, that is
$$\int_a^b f(x)\mathrm{d}x=\lim_{\lambda\to 0}\sum_{i=1}^{n}f(\xi_i)\Delta x_i,$$
where $f(x)$ is called the **integrand function**, $f(x)\mathrm{d}x$ is called the **integrand expression**, x is called the **integral variable**, a is called the **lower limit of integral**, b is called the **upper limit of integral**, $[a,b]$ is called the **integral interval**.

The sum $\sum_{i=1}^{n}f(\xi_i)\Delta x_i$ is called the **definite integral sum** of $f(x)$. If the function $f(x)$ have a definite integral on $[a,b]$, then we say that $f(x)$ is **integrable** on $[a,b]$.

According to the definition of definite integral, the area of the trapezoid with curve side is
$$S=\int_a^b f(x)\mathrm{d}x,$$
the distance of variable speed liner motion is
$$s=\int_{T_1}^{T_2}v(t)\mathrm{d}t.$$

定义 6.1 设函数 $f(x)$ 在 $[a,b]$ 上有界. 在 $[a,b]$ 中任意插入 $n-1$ 个分点
$$a=x_0<x_1<x_2<\cdots<x_{n-1}<x_n=b,$$
把区间 $[a,b]$ 分成 n 个小区间:
$$[x_0,x_1], [x_1,x_2], \cdots, [x_{n-1},x_n].$$
各小段区间的长依次为
$$\Delta x_1=x_1-x_0, \Delta x_2=x_2-x_1,$$
$$\cdots, \Delta x_n=x_n-x_{n-1}.$$
在每个小区间 $[x_{i-1},x_i]$ 上任取一个点 ξ_i, 作函数值 $f(\xi_i)$ 与小区间长度 Δx_i 的乘积
$$f(\xi_i)\Delta x_i \quad (i=1,2,\cdots,n),$$
并作出和式
$$A=\sum_{i=1}^{n}f(\xi_i)\Delta x_i.$$
记 $\lambda=\max\{\Delta x_1,\Delta x_2,\cdots,\Delta x_n\}$. 如果不论如何划分 $[a,b]$, 也不论在小区间 $[x_{i-1},x_i]$ 上如何选取点 ξ_i, 当 $\lambda \to 0$ 时, 和式 A 总趋于确定的极限 I, 那么称这个极限 I 为函数 $f(x)$ 在区间 $[a,b]$ 上的**定积分**, 记作 $\int_a^b f(x)\mathrm{d}x$, 即
$$\int_a^b f(x)\mathrm{d}x=\lim_{\lambda\to 0}\sum_{i=1}^{n}f(\xi_i)\Delta x_i,$$
其中 $f(x)$ 叫作被积函数, $f(x)\mathrm{d}x$ 叫作被积表达式, x 叫作积分变量, a 叫作积分下限, b 叫作积分上限, $[a,b]$ 叫作积分区间.

和式 $\sum_{i=1}^{n}f(\xi_i)\Delta x_i$ 通常称为 $f(x)$ 的积分和. 如果函数 $f(x)$ 在 $[a,b]$ 上的定积分存在, 我们就说 $f(x)$ 在区间 $[a,b]$ 上**可积**.

根据定积分的定义, 曲边梯形的面积为
$$S=\int_a^b f(x)\mathrm{d}x,$$
变速直线运动的路程为
$$s=\int_{T_1}^{T_2}v(t)\mathrm{d}t.$$

Remark The value of definite integral is just connected with the integrand function and the integral interval, but the integral variable, that is
$$\int_a^b f(x)\mathrm{d}x = \int_a^b f(t)\mathrm{d}t = \int_a^b f(u)\mathrm{d}u.$$

If the function $f(x)$ is integrable on $[a,b]$, which conditions should it satisfy on $[a,b]$?

Theorem 6.1 If the function $f(x)$ is contiuous on interval $[a,b]$, then $f(x)$ is integrable on $[a,b]$.

Theorem 6.2 If the function $f(x)$ is bounded on interval $[a,b]$, and it has finite discontinuous points, then $f(x)$ is integrable on $[a,b]$.

3. The Geometric Significance of the Definite Integral

On the interval $[a,b]$, if $f(x) \geqslant 0$, in geometry, definite integral $\int_a^b f(x)\mathrm{d}x$ equals to the area of the trapezoid with curve side which enclosed by the curve $y=f(x)$, the line $x=a$ and $x=b$, and x-axis.

If $f(x) \leqslant 0$, the trapezoid with curve side which enclosed by the curve $y=f(x)$, the line $x=a$ and $x=b$, and x-axis located under x axis, in geometry, the definite integral $\int_a^b f(x)\mathrm{d}x$ equals to the negative of the area of the trapezoid with curve side enclosed above. In fact,
$$\begin{aligned}\int_a^b f(x)\mathrm{d}x &= \lim_{\lambda \to 0}\sum_{i=1}^n f(\xi_i)\Delta x_i \\ &= -\lim_{\lambda \to 0}\sum_{i=1}^n [-f(\xi_i)]\Delta x_i \\ &= -\int_a^b [-f(x)]\mathrm{d}x.\end{aligned}$$

If on the interval $[a,b]$, $f(x)$ takes both positive and negative values, that is, some graph of $f(x)$ located above x-axis, others located under x-axis. And we give the area a sign, the area which graph located above x-axis is " $+$ ", and the area which graph located below x-axis is " $-$ ", in geometry, the definite integral $\int_a^b f(x)\mathrm{d}x$ equals to the algebra sum area of the graph which enclosed by the curve $y=f(x)$, the line $x=a$ and $x=b$, and x-axis, see Figure 6.2.

注 定积分的值只与被积函数及积分区间有关,而与积分变量的记法无关,即
$$\int_a^b f(x)\mathrm{d}x = \int_a^b f(t)\mathrm{d}t = \int_a^b f(u)\mathrm{d}u.$$

函数 $f(x)$ 在 $[a,b]$ 上满足什么条件时,它在 $[a,b]$ 上可积呢?

定理 6.1 如果函数 $f(x)$ 在区间 $[a,b]$ 上连续,则 $f(x)$ 在 $[a,b]$ 上可积.

定理 6.2 如果函数 $f(x)$ 在区间 $[a,b]$ 上有界,且只有有限个间断点,则 $f(x)$ 在 $[a,b]$ 上可积.

3. 定积分的几何意义

在区间 $[a,b]$ 上,如果 $f(x) \geqslant 0$,定积分 $\int_a^b f(x)\mathrm{d}x$ 在几何上表示由曲线 $y=f(x)$ 和直线 $x=a,x=b$ 及 x 轴所围成的曲边梯形的面积.

如果 $f(x) \leqslant 0$,由曲线 $y=f(x)$ 和直线 $x=a,x=b$ 及 x 轴所围成的曲边梯形位于 x 轴的下方,定积分 $\int_a^b f(x)\mathrm{d}x$ 在几何上表示该曲边梯形面积的负值.事实上,
$$\begin{aligned}\int_a^b f(x)\mathrm{d}x &= \lim_{\lambda \to 0}\sum_{i=1}^n f(\xi_i)\Delta x_i \\ &= -\lim_{\lambda \to 0}\sum_{i=1}^n [-f(\xi_i)]\Delta x_i \\ &= -\int_a^b [-f(x)]\mathrm{d}x.\end{aligned}$$

如果 $f(x)$ 在区间 $[a,b]$ 上既取得正值又取得负值时,即函数 $f(x)$ 的图像的某些部分在 x 轴的上方,而其他部分在 x 轴的下方.如果我们对面积赋以正、负号,在 x 轴上方的图像面积赋以正号,在 x 轴下方的图形面积赋以负号,则定积分 $\int_a^b f(x)\mathrm{d}x$ 在几何上表示介于 x 轴、函数 $f(x)$ 的图像及两条直线 $x=a,x=b$ 之间的各部分面积的代数和,见图 6.2.

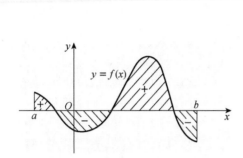

Figure 6.2
图 6.2

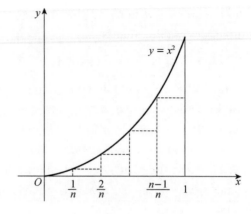

Figure 6.3
图 6.3

Example 1 Compute $\int_0^1 x^2 \, dx$ by the definition of definite integral.

Solution As shown in Figure 6.3, divide the interval $[0,1]$ into n equal parts, the dividing points are

$$x_i = \frac{i}{n} \quad (i=1,2,\cdots,n-1),$$

the length of sub-intervals are

$$\Delta x_i = \frac{1}{n} \quad (i=1,2,\cdots,n).$$

Take $\xi_i = \frac{i}{n}(i=1,2,\cdots,n)$, compute

$$\begin{aligned}\sum_{i=1}^n f(\xi_i)\Delta x_i &= \sum_{i=1}^n \xi_i^2 \Delta x_i \\ &= \sum_{i=1}^n \left(\frac{i}{n}\right)^2 \frac{1}{n} = \frac{1}{n^3}\sum_{i=1}^n i^2 \\ &= \frac{1}{n^3} \cdot \frac{1}{6}n(n+1)(2n+1) \\ &= \frac{1}{6}\left(1+\frac{1}{n}\right)\left(2+\frac{1}{n}\right).\end{aligned}$$

As $\lambda = \frac{1}{n}$, if $\lambda \to 0$, $n \to \infty$, so

$$\begin{aligned}\int_0^1 x^2 \, dx &= \lim_{\lambda\to 0}\sum_{i=1}^n f(\xi_i)\Delta x_i \\ &= \lim_{n\to\infty}\frac{1}{6}\left(1+\frac{1}{n}\right)\left(2+\frac{1}{n}\right) = \frac{1}{3}.\end{aligned}$$

例 1 利用定积分的定义计算 $\int_0^1 x^2 \, dx$.

解 如图 6.3 所示，把区间 $[0,1]$ 分成 n 等份，分点为

$$x_i = \frac{i}{n} \quad (i=1,2,\cdots,n-1),$$

小区间长度为

$$\Delta x_i = \frac{1}{n} \quad (i=1,2,\cdots,n).$$

取 $\xi_i = \frac{i}{n}(i=1,2,\cdots,n)$，作积分和

$$\begin{aligned}\sum_{i=1}^n f(\xi_i)\Delta x_i &= \sum_{i=1}^n \xi_i^2 \Delta x_i \\ &= \sum_{i=1}^n \left(\frac{i}{n}\right)^2 \frac{1}{n} = \frac{1}{n^3}\sum_{i=1}^n i^2 \\ &= \frac{1}{n^3} \cdot \frac{1}{6}n(n+1)(2n+1) \\ &= \frac{1}{6}\left(1+\frac{1}{n}\right)\left(2+\frac{1}{n}\right).\end{aligned}$$

因为 $\lambda = \frac{1}{n}$，当 $\lambda \to 0$ 时，$n \to \infty$，所以

$$\begin{aligned}\int_0^1 x^2 \, dx &= \lim_{\lambda\to 0}\sum_{i=1}^n f(\xi_i)\Delta x_i \\ &= \lim_{n\to\infty}\frac{1}{6}\left(1+\frac{1}{n}\right)\left(2+\frac{1}{n}\right) = \frac{1}{3}.\end{aligned}$$

Example 2 Compute
$$\int_0^1 (1-x)\,dx$$
by geometrical significance of definite integral.

Solution Definite integral of the function $y=1-x$ on interval $[0,1]$ is the area of the trapezoid with curve bottomed by the interval $[0,1]$. The trapezoid with curve is a right triangle which is enclosed by the interval $[0,1]$ and $y=1-x$, its length of side and hight are both 1 (Figure 6.4), so
$$\int_0^1 (1-x)\,dx = \frac{1}{2} \times 1 \times 1 = \frac{1}{2}.$$

例 2 利用定积分的几何意义求
$$\int_0^1 (1-x)\,dx.$$

解 函数 $y=1-x$ 在区间 $[0,1]$ 上的定积分是以区间 $[0,1]$ 为底的曲边梯形的面积. 因为曲边梯形是以 $y=1-x$ 为曲边, 以区间 $[0,1]$ 为底的直角三角形, 其底边长及高均为 1(图 6.4), 所以
$$\int_0^1 (1-x)\,dx = \frac{1}{2} \times 1 \times 1 = \frac{1}{2}.$$

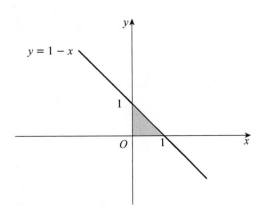

Figure 6.4
图 6.4

4. The Properties of the Definite Integral

Two rules:

(1) $\int_a^b f(x)\,dx = 0$, when $a=b$;

(2) $\int_a^b f(x)\,dx = -\int_b^a f(x)\,dx$, when $a>b$.

Remark In the following properties, the limits of integral are without limitation except specialized.

Property 1 The definite integral of the sum(difference) of two functions is equal to the sum(difference) of the definite integral of every function, or
$$\int_a^b [f(x) \pm g(x)]\,dx$$
$$= \int_a^b f(x)\,dx \pm \int_a^b g(x)\,dx.$$

4. 定积分的性质

两点规定:

(1) 当 $a=b$ 时, $\int_a^b f(x)\,dx = 0$;

(2) 当 $a>b$ 时, $\int_a^b f(x)\,dx = -\int_b^a f(x)\,dx$.

注 下列性质中如不对积分上、下限的大小特别指明, 则表示不加限制.

性质 1 两个函数的和(差)的定积分等于两个函数的定积分的和(差), 即
$$\int_a^b [f(x) \pm g(x)]\,dx$$
$$= \int_a^b f(x)\,dx \pm \int_a^b g(x)\,dx.$$

Property 2 The constant factor of the integrand function could be move to the out of the sign of integration, or
$$\int_a^b kf(x)\mathrm{d}x = k\int_a^b f(x)\mathrm{d}x.$$

Property 3 If the integral interval is divided into two sub-integrals, then the definite integral on the whole interval equals to the sum of the definite integral on every sub-interval, or
$$\int_a^b f(x)\mathrm{d}x = \int_a^c f(x)\mathrm{d}x + \int_c^b f(x)\mathrm{d}x.$$

This property indicate that the definite integral is additive about integral interval. It's worth noting that we always have
$$\int_a^b f(x)\mathrm{d}x = \int_a^c f(x)\mathrm{d}x + \int_c^b f(x)\mathrm{d}x$$
regardless of the size of a, b, c. For example, if $a<b<c$, as
$$\int_a^c f(x)\mathrm{d}x = \int_a^b f(x)\mathrm{d}x + \int_b^c f(x)\mathrm{d}x,$$
so
$$\int_a^b f(x)\mathrm{d}x = \int_a^c f(x)\mathrm{d}x - \int_b^c f(x)\mathrm{d}x$$
$$= \int_a^c f(x)\mathrm{d}x + \int_c^b f(x)\mathrm{d}x.$$

Property 4 If $f(x)\equiv 1$ on the interval $[a,b]$, then
$$\int_a^b 1\mathrm{d}x = \int_a^b \mathrm{d}x = b-a.$$

Property 5 If $f(x)\geqslant 0$ on the interval $[a,b]$, then
$$\int_a^b f(x)\mathrm{d}x \geqslant 0.$$

Corollary 1 If $f(x)\leqslant g(x)$ on the interval $[a,b]$, then
$$\int_a^b f(x)\mathrm{d}x \leqslant \int_a^b g(x)\mathrm{d}x.$$

In fact, as $g(x)-f(x)\geqslant 0$, then
$$\int_a^b g(x)\mathrm{d}x - \int_a^b f(x)\mathrm{d}x$$
$$= \int_a^b [g(x)-f(x)]\mathrm{d}x \geqslant 0,$$
so
$$\int_a^b f(x)\mathrm{d}x \leqslant \int_a^b g(x)\mathrm{d}x.$$

Corollary 2 $\left|\int_a^b f(x)\mathrm{d}x\right| \leqslant \int_a^b |f(x)|\mathrm{d}x \ (a<b).$

性质 2 被积函数的常数因子可以提到积分号外面,即
$$\int_a^b kf(x)\mathrm{d}x = k\int_a^b f(x)\mathrm{d}x.$$

性质 3 如果将积分区间分成两部分,则在整个区间上的定积分等于这两部分区间上定积分之和,即
$$\int_a^b f(x)\mathrm{d}x = \int_a^c f(x)\mathrm{d}x + \int_c^b f(x)\mathrm{d}x.$$

这个性质表明定积分对于积分区间具有可加性.值得注意的是,对于 a,b,c 的任何大小顺序,总有等式
$$\int_a^b f(x)\mathrm{d}x = \int_a^c f(x)\mathrm{d}x + \int_c^b f(x)\mathrm{d}x.$$
例如,当 $a<b<c$ 时,由于
$$\int_a^c f(x)\mathrm{d}x = \int_a^b f(x)\mathrm{d}x + \int_b^c f(x)\mathrm{d}x,$$
于是有
$$\int_a^b f(x)\mathrm{d}x = \int_a^c f(x)\mathrm{d}x - \int_b^c f(x)\mathrm{d}x$$
$$= \int_a^c f(x)\mathrm{d}x + \int_c^b f(x)\mathrm{d}x.$$

性质 4 如果在区间 $[a,b]$ 上 $f(x)\equiv 1$,则
$$\int_a^b 1\mathrm{d}x = \int_a^b \mathrm{d}x = b-a.$$

性质 5 如果在区间 $[a,b]$ 上 $f(x)\geqslant 0$,则
$$\int_a^b f(x)\mathrm{d}x \geqslant 0.$$

推论 1 如果在区间 $[a,b]$ 上 $f(x)\leqslant g(x)$,则
$$\int_a^b f(x)\mathrm{d}x \leqslant \int_a^b g(x)\mathrm{d}x.$$

事实上,因为 $g(x)-f(x)\geqslant 0$,从而
$$\int_a^b g(x)\mathrm{d}x - \int_a^b f(x)\mathrm{d}x$$
$$= \int_a^b [g(x)-f(x)]\mathrm{d}x \geqslant 0,$$
所以 $\int_a^b f(x)\mathrm{d}x \leqslant \int_a^b g(x)\mathrm{d}x.$

推论 2 $\left|\int_a^b f(x)\mathrm{d}x\right| \leqslant \int_a^b |f(x)|\mathrm{d}x \ (a<b).$

In fact, as $-|f(x)| \leqslant f(x) \leqslant |f(x)|$, so
$$-\int_a^b |f(x)|\,dx \leqslant \int_a^b f(x)\,dx$$
$$\leqslant \int_a^b |f(x)|\,dx,$$
or
$$\left|\int_a^b f(x)\,dx\right| \leqslant \int_a^b |f(x)|\,dx.$$

Property 6 If M and m are maximum and minimum of $f(x)$ on the interval $[a,b]$ respectively, then
$$m(b-a) \leqslant \int_a^b f(x)\,dx \leqslant M(b-a).$$

Proof As $m \leqslant f(x) \leqslant M$, so
$$\int_a^b m\,dx \leqslant \int_a^b f(x)\,dx \leqslant \int_a^b M\,dx,$$
then
$$m(b-a) \leqslant \int_a^b f(x)\,dx \leqslant M(b-a).$$

Property 7 (The Mean Value Theorem of Definite Integral)
If the function $f(x)$ is continuous on the closed interval $[a,b]$, then exist at least one point ξ in the integral interval $[a,b]$, such that
$$\int_a^b f(x)\,dx = f(\xi)(b-a) \quad (a \leqslant \xi \leqslant b).$$

This is called the **mean value formula of definite integral**.

Proof From Property 6, we know
$$m(b-a) \leqslant \int_a^b f(x)\,dx \leqslant M(b-a).$$

Divided by $b-a$, so
$$m \leqslant \frac{1}{b-a}\int_a^b f(x)\,dx \leqslant M.$$

By the intermediate value theorem of continuous function, exist more than one point ξ in $[a,b]$, such that
$$f(\xi) = \frac{1}{b-a}\int_a^b f(x)\,dx.$$

Then multiplied by $b-a$, so
$$\int_a^b f(x)\,dx = f(\xi)(b-a).$$

Remark Whether $a<b$ or $a>b$, the mean value formula of definite integral is always right. $f(\xi) = \frac{1}{b-a}\int_a^b f(x)\,dx$ is called the **mean value** of $f(x)$ on the closed interval $[a,b]$.

事实上,因为 $-|f(x)| \leqslant f(x) \leqslant |f(x)|$,所以
$$-\int_a^b |f(x)|\,dx \leqslant \int_a^b f(x)\,dx$$
$$\leqslant \int_a^b |f(x)|\,dx,$$
即
$$\left|\int_a^b f(x)\,dx\right| \leqslant \int_a^b |f(x)|\,dx.$$

性质 6 设 M 及 m 分别是函数 $f(x)$ 在区间 $[a,b]$ 上的最大值及最小值,则
$$m(b-a) \leqslant \int_a^b f(x)\,dx \leqslant M(b-a).$$

证明 因为 $m \leqslant f(x) \leqslant M$,所以
$$\int_a^b m\,dx \leqslant \int_a^b f(x)\,dx \leqslant \int_a^b M\,dx,$$
从而
$$m(b-a) \leqslant \int_a^b f(x)\,dx \leqslant M(b-a).$$

性质 7(定积分中值定理)
如果函数 $f(x)$ 在闭区间 $[a,b]$ 上连续,则在积分区间 $[a,b]$ 上至少存在一个点 ξ,使得
$$\int_a^b f(x)\,dx = f(\xi)(b-a) \quad (a \leqslant \xi \leqslant b).$$

这个公式叫作**积分中值公式**.

证明 由性质 6 可知
$$m(b-a) \leqslant \int_a^b f(x)\,dx \leqslant M(b-a).$$

各项除以 $b-a$,得
$$m \leqslant \frac{1}{b-a}\int_a^b f(x)\,dx \leqslant M.$$

由连续函数的介值定理,在 $[a,b]$ 上至少存在一点 ξ,使得
$$f(\xi) = \frac{1}{b-a}\int_a^b f(x)\,dx.$$

再两端乘以 $b-a$,得
$$\int_a^b f(x)\,dx = f(\xi)(b-a).$$

注 不论 $a<b$ 还是 $a>b$,积分中值公式都成立. $f(\xi) = \frac{1}{b-a}\int_a^b f(x)\,dx$ 称为 $f(x)$ 在闭区间 $[a,b]$ 上的**平均值**.

6.2 The Fundamental Formula of Calculus
6.2 微积分基本公式

1. The Function of Integral Upper Limit and Its Derivative

Suppose the function $f(x)$ is continuous on the interval $[a,b]$, and x is a point in $[a,b]$. $\int_a^x f(x)dx$, the definite integral of the function $f(x)$ on the sub-interval $[a,x]$, is called the **function of integral upper limit**, which is defined on the interval $[a,b]$, signed as

$$\Phi(x) = \int_a^x f(x)dx,$$

or

$$\Phi(x) = \int_a^x f(t)dt.$$

Theorem 6.3 If the function $f(x)$ is continuous on the interval $[a,b]$, then the function of integral upper limit, $\Phi(x) = \int_a^x f(t)dt$, have derivative on $[a,b]$, and

$$\Phi'(x) = \frac{d}{dx}\int_a^x f(t)dt = f(x) \quad (a \leqslant x \leqslant b).$$

Theorem 6.4 If the function $f(x)$ is continuous on the interval $[a,b]$, then the function $\Phi(x) = \int_a^x f(t)dt$ is a primitive function of $f(x)$ on $[a,b]$.

The significance of Theorem 6.4: It sure that the primitive function of a continuous function exists, and uncover the relationship between definite integral and primitive function in integral calculus.

2. The Newton-Leibniz Formula

We already know the relation between the distance function and the velocity function in a variable linear motion. If the object move rectilinear from some fixed point, the distance is $s(t)$ at t, the speed is

$$v = v(t) = s'(t) \quad (v(t) \geqslant 0),$$

so the distance s in the time interval $[T_1, T_2]$ could be expressed as

$$s(T_2) - s(T_1) \quad \text{or} \quad \int_{T_1}^{T_2} v(t)dt,$$

1. 积分上限函数及其导数

设函数 $f(x)$ 在区间 $[a,b]$ 上连续,并且设 x 为 $[a,b]$ 上的一点. 我们把函数 $f(x)$ 在部分区间 $[a,x]$ 上的定积分 $\int_a^x f(x)dx$ 称为**积分上限函数**. 它是区间 $[a,b]$ 上的函数,记为

$$\Phi(x) = \int_a^x f(x)dx$$

或

$$\Phi(x) = \int_a^x f(t)dt.$$

定理 6.3 如果函数 $f(x)$ 在区间 $[a,b]$ 上连续,则积分上限函数 $\Phi(x) = \int_a^x f(t)dt$ 在 $[a,b]$ 上具有导数,并且

$$\Phi'(x) = \frac{d}{dx}\int_a^x f(t)dt = f(x) \quad (a \leqslant x \leqslant b).$$

定理 6.4 如果函数 $f(x)$ 在区间 $[a,b]$ 上连续,则函数 $\Phi(x) = \int_a^x f(t)dt$ 就是 $f(x)$ 在 $[a,b]$ 上的一个原函数.

定理 6.4 的重要意义:它肯定了连续函数的原函数是存在的,并且初步揭示了积分学中的定积分与原函数之间的联系.

2. 牛顿-莱布尼茨公式

我们已经知道变速直线运动中路程函数与速度函数之间的联系. 设物体从某定点开始做直线运动,到 t 时刻,所经过的路程为 $s(t)$,速度为

$$v = v(t) = s'(t) \quad (v(t) \geqslant 0),$$

则在时间段 $[T_1, T_2]$ 内物体所经过的路程 s 可表示为

$$s(T_2) - s(T_1) \quad \text{或} \quad \int_{T_1}^{T_2} v(t)dt,$$

or
$$\int_{T_1}^{T_2} v(t)dt = s(T_2) - s(T_1).$$

This shows that the definite integral of the speed function $v(t)$ on the interval $[T_1, T_2]$ equals to the increment of $s(t)$, or the primitive function of $v(t)$, in interval $[T_1, T_2]$.

Is the relationship universal which is concluded from this special problem? The answer is yes. It is the famous Newton-Leibniz formula.

Theorem 6.5 If $F(x)$ is a primitive function of the continuous function $f(x)$ on the interval $[a,b]$, then
$$\int_a^b f(x)dx = F(b) - F(a).$$

This is called the **Newton-Leibniz formula**, also as the **fundamental formula of calculus**.

Proof As $F(x)$ is a primitive function of the continuous function $f(x)$, according to Theorem 6.4, the function of integral upper limit, $\Phi(x) = \int_a^x f(t)dt$, is also a primitive function of $f(x)$, then exist a constant C, make
$$F(x) - \Phi(x) = C \quad (a \leqslant x \leqslant b).$$

When $x = a$, then
$$F(a) - \Phi(a) = C,$$
and $\Phi(a) = 0$, so $C = F(a)$.

When $x = b$, then
$$F(b) - \Phi(b) = F(a),$$
so $\Phi(b) = F(b) - F(a)$, or
$$\int_a^b f(x)dx = F(b) - F(a).$$

For simplicity, sign $F(b) - F(a)$ as $[F(x)]_a^b$, then
$$\int_a^b f(x)dx = [F(x)]_a^b$$
$$= F(b) - F(a).$$

The formula uncovered the relationship between definite integral and primitive function or indefinite integral again.

Example 1 Compute $\int_0^1 x^2 dx$.

Solution As $\frac{1}{3}x^3$ is a primitive function of x^2, so
$$\int_0^1 x^2 dx = \left[\frac{1}{3}x^3\right]_0^1$$

即
$$\int_{T_1}^{T_2} v(t)dt = s(T_2) - s(T_1).$$

这表明,速度函数 $v(t)$ 在区间 $[T_1, T_2]$ 上的定积分等于 $v(t)$ 的原函数 $s(t)$ 在 $[T_1, T_2]$ 上的增量.

从这个特殊问题中得出的关系是否具有普遍意义呢?答案是肯定的,它就是著名的牛顿-莱布尼茨公式.

定理 6.5 如果 $F(x)$ 是连续函数 $f(x)$ 在区间 $[a,b]$ 上的一个原函数,则
$$\int_a^b f(x)dx = F(b) - F(a).$$

此公式称为**牛顿-莱布尼茨公式**,也称为**微积分基本公式**.

证明 由于 $F(x)$ 是连续函数 $f(x)$ 的一个原函数,又根据定理 6.4,积分上限函数 $\Phi(x) = \int_a^x f(t)dt$ 也是 $f(x)$ 的一个原函数,于是存在一个常数 C,使得
$$F(x) - \Phi(x) = C \quad (a \leqslant x \leqslant b).$$

当 $x = a$ 时,有
$$F(a) - \Phi(a) = C,$$
而 $\Phi(a) = 0$,所以 $C = F(a)$.

当 $x = b$ 时,有
$$F(b) - \Phi(b) = F(a),$$
所以 $\Phi(b) = F(b) - F(a)$,即
$$\int_a^b f(x)dx = F(b) - F(a).$$

为了方便起见,可把 $F(b) - F(a)$ 记成 $[F(x)]_a^b$,于是
$$\int_a^b f(x)dx = [F(x)]_a^b$$
$$= F(b) - F(a).$$

这一公式进一步揭示了定积分与被积函数的原函数或不定积分之间的联系.

例 1 计算 $\int_0^1 x^2 dx$.

解 由于 $\frac{1}{3}x^3$ 是 x^2 的一个原函数,所以
$$\int_0^1 x^2 dx = \left[\frac{1}{3}x^3\right]_0^1$$

$$= \frac{1}{3} \cdot 1^3 - \frac{1}{3} \cdot 0^3 = \frac{1}{3}.$$

Example 2 Compute $\int_{-1}^{\sqrt{3}} \frac{dx}{1+x^2}$.

Solution As $\arctan x$ is a primitive function of $\frac{1}{1+x^2}$, so

$$\int_{-1}^{\sqrt{3}} \frac{dx}{1+x^2} = [\arctan x]_{-1}^{\sqrt{3}}$$
$$= \arctan\sqrt{3} - \arctan(-1)$$
$$= \frac{\pi}{3} - \left(-\frac{\pi}{4}\right) = \frac{7}{12}\pi.$$

Example 3 Compute $\int_{-2}^{-1} \frac{1}{x} dx$.

Solution $\int_{-2}^{-1} \frac{1}{x} dx = [\ln|x|]_{-2}^{-1} = \ln 1 - \ln 2$
$$= -\ln 2.$$

Example 4 Compute the area of the graph which enclosed by the cine curve $y = \sin x$ on $[0, \pi]$ and x-axis.

Solution The graph is a special case of trapezoid with curve side. The area is

$$S = \int_0^\pi \sin x \, dx = [-\cos x]_0^\pi$$
$$= -(-1-1) = 2.$$

Example 5 A car moves forward as 36 km/h, some time later will slow down and stop. If the car's accelerated speed is $a = -5$ m/s^2, how long does it move from breaking to stop?

Solution When $t = 0$, the car's speed is

$$v_0 = 36 \text{ km/h} = \frac{36 \times 1000}{3600} \text{m/s} = 10 \text{ m/s}.$$

Its speed in time t is

$$v(t) = v_0 + at = 10 - 5t.$$

When the car stops, its speed is $v(t) = 0$. From $v(t) = 10 - 5t = 0$, then $t = 2$. This is the time from breaking to stop. So the distance is

$$s = \int_0^2 v(t) dt = \int_0^2 (10 - 5t) dt$$
$$= \left[10t - 5 \cdot \frac{1}{2} t^2\right]_0^2 = 10,$$

or the car should move 10 meters from breaking to stop.

Example 6 Compute $\lim\limits_{x \to 0} \dfrac{\int_{\cos x}^{1} e^{-t^2} dt}{x^2}$.

Solution This is an indeterminate form $\dfrac{0}{0}$. From the L'Hospital Rule, we have

$$\lim_{x \to 0} \frac{\int_{\cos x}^{1} e^{-t^2} dt}{x^2} = \lim_{x \to 0} \frac{-\int_{1}^{\cos x} e^{-t^2} dt}{x^2}$$
$$= \lim_{x \to 0} \frac{\sin x e^{-\cos^2 x}}{2x} = \frac{1}{2e}.$$

Guide If $\Phi(x) = \int_{1}^{x} e^{-t^2} dt$, then

$$\Phi(\cos x) = \int_{1}^{\cos x} e^{-t^2} dt,$$

so

$$\frac{d}{dx} \int_{1}^{\cos x} e^{-t^2} dt = \frac{d}{dx} \Phi(\cos x)$$
$$\xrightarrow{\text{Let } u = \cos x} \frac{d}{du} \Phi(u) \cdot \frac{du}{dx}$$
$$= e^{-u^2} \cdot (-\sin x)$$
$$= -\sin x \cdot e^{-\cos^2 x}.$$

6.3 Definite Integration by Substitution and Parts

1. Definite Integration by Substitution

Theorem 6.6 If the function $f(x)$ is continuous on the interval $[a, b]$, the function $x = \varphi(t)$ satisfies:

(1) $\varphi(\alpha) = a$, $\varphi(\beta) = b$;

(2) $\varphi(t)$ exists continuous derivative on $[\alpha, \beta]$ (or $[\beta, \alpha]$), and its range is not beyond $[a, b]$,

then

$$\int_{a}^{b} f(x) dx = \int_{\alpha}^{\beta} f(\varphi(t)) \varphi'(t) dt.$$

This is called the **formula for definite integration by substitution**.

Remark (1) When we replace x by $\varphi(t)$, the limits of integration are also replaced by t.

例 6 计算 $\lim\limits_{x \to 0} \dfrac{\int_{\cos x}^{1} e^{-t^2} dt}{x^2}$.

解 这是一个 $\dfrac{0}{0}$ 型未定式. 由洛必达法则有

$$\lim_{x \to 0} \frac{\int_{\cos x}^{1} e^{-t^2} dt}{x^2} = \lim_{x \to 0} \frac{-\int_{1}^{\cos x} e^{-t^2} dt}{x^2}$$
$$= \lim_{x \to 0} \frac{\sin x e^{-\cos^2 x}}{2x} = \frac{1}{2e}.$$

提示 设 $\Phi(x) = \int_{1}^{x} e^{-t^2} dt$, 则

$$\Phi(\cos x) = \int_{1}^{\cos x} e^{-t^2} dt.$$

于是

$$\frac{d}{dx} \int_{1}^{\cos x} e^{-t^2} dt = \frac{d}{dx} \Phi(\cos x)$$
$$\xrightarrow{\text{令 } u = \cos x} \frac{d}{du} \Phi(u) \cdot \frac{du}{dx}$$
$$= e^{-u^2} \cdot (-\sin x)$$
$$= -\sin x \cdot e^{-\cos^2 x}.$$

6.3 定积分的换元法和分部积分法

1. 定积分的换元法

定理 6.6 假设函数 $f(x)$ 在区间 $[a, b]$ 上连续, 函数 $x = \varphi(t)$ 满足条件:

(1) $\varphi(\alpha) = a$, $\varphi(\beta) = b$;

(2) $\varphi(t)$ 在 $[\alpha, \beta]$ (或 $[\beta, \alpha]$) 上具有连续导数, 且其值域不越出 $[a, b]$,

则有

$$\int_{a}^{b} f(x) dx = \int_{\alpha}^{\beta} f(\varphi(t)) \varphi'(t) dt.$$

这个公式叫作**定积分的换元公式**.

注 (1) 作替换 $x = \varphi(t)$ 时, 积分限也要换成变量 t 的积分限;

(2) After computed the primitive function of $f(\varphi(t))\varphi'(t)$, we don't need to revert it to function of x, but compute the difference value of limits of integration.

Example 1 Compute $\int_0^a \sqrt{a^2-x^2}\,dx\ (a>0)$.

Solution Let $x=a\sin t$. If $x=0$, then $t=0$; if $x=a$, then $t=\frac{\pi}{2}$. So

$$\int_0^a \sqrt{a^2-x^2}\,dx = \int_0^{\pi/2} a\cos t \cdot a\cos t\,dt$$
$$= a^2 \int_0^{\pi/2} \cos^2 t\,dt = \frac{a^2}{2}\int_0^{\pi/2}(1+\cos 2t)\,dt$$
$$= \frac{a^2}{2}\left[t+\frac{1}{2}\sin 2t\right]_0^{\pi/2} = \frac{1}{4}\pi a^2.$$

Example 2 Compute $\int_0^{\pi/2}\cos^5 x\sin x\,dx$.

Solution Let $t=\cos x$. If $x=0$, then $t=1$; if $x=\frac{\pi}{2}$, then $t=0$. So

$$\int_0^{\pi/2}\cos^5 x\sin x\,dx = -\int_0^{\pi/2}\cos^5 x\,d(\cos x)$$
$$= -\int_1^0 t^5\,dt = \int_0^1 t^5\,dt = \left[\frac{1}{6}t^6\right]_0^1 = \frac{1}{6}.$$

When we don't write the variable t, the limits of integration are remain unchanged:

$$\int_0^{\pi/2}\cos^5 x\sin x\,dx = -\int_0^{\pi/2}\cos^5 x\,d(\cos x)$$
$$= -\left[\frac{1}{6}\cos^6 x\right]_0^{\pi/2}$$
$$= -\frac{1}{6}\cos^6\frac{\pi}{2} + \frac{1}{6}\cos^6 0$$
$$= \frac{1}{6}.$$

Example 3 Compute $\int_0^{\pi}\sqrt{\sin^3 x - \sin^5 x}\,dx$.

Solution

$$\int_0^{\pi}\sqrt{\sin^3 x - \sin^5 x}\,dx = \int_0^{\pi}\sin^{3/2}x|\cos x|\,dx$$
$$= \int_0^{\pi/2}\sin^{3/2}x\cos x\,dx - \int_{\pi/2}^{\pi}\sin^{3/2}x\cos x\,dx$$

（2）求出 $f(\varphi(t))\varphi'(t)$ 的原函数后，不必还原为 x 的函数，只要代入 t 的上、下限求差值即可。

例 1 计算 $\int_0^a \sqrt{a^2-x^2}\,dx\ (a>0)$.

解 令 $x=a\sin t$. 当 $x=0$ 时 $t=0$；当 $x=a$ 时 $t=\frac{\pi}{2}$. 所以

$$\int_0^a \sqrt{a^2-x^2}\,dx = \int_0^{\pi/2} a\cos t \cdot a\cos t\,dt$$
$$= a^2 \int_0^{\pi/2} \cos^2 t\,dt = \frac{a^2}{2}\int_0^{\pi/2}(1+\cos 2t)\,dt$$
$$= \frac{a^2}{2}\left[t+\frac{1}{2}\sin 2t\right]_0^{\pi/2} = \frac{1}{4}\pi a^2.$$

例 2 计算 $\int_0^{\pi/2}\cos^5 x\sin x\,dx$.

解 令 $t=\cos x$. 当 $x=0$ 时，$t=1$；当 $x=\frac{\pi}{2}$ 时，$t=0$. 所以

$$\int_0^{\pi/2}\cos^5 x\sin x\,dx = -\int_0^{\pi/2}\cos^5 x\,d(\cos x)$$
$$= -\int_1^0 t^5\,dt = \int_0^1 t^5\,dt = \left[\frac{1}{6}t^6\right]_0^1 = \frac{1}{6}.$$

当不写出变量 t 时，积分上下限不变：

$$\int_0^{\pi/2}\cos^5 x\sin x\,dx = -\int_0^{\pi/2}\cos^5 x\,d(\cos x)$$
$$= -\left[\frac{1}{6}\cos^6 x\right]_0^{\pi/2}$$
$$= -\frac{1}{6}\cos^6\frac{\pi}{2} + \frac{1}{6}\cos^6 0$$
$$= \frac{1}{6}.$$

例 3 计算 $\int_0^{\pi}\sqrt{\sin^3 x - \sin^5 x}\,dx$.

解

$$\int_0^{\pi}\sqrt{\sin^3 x - \sin^5 x}\,dx = \int_0^{\pi}\sin^{3/2}x|\cos x|\,dx$$
$$= \int_0^{\pi/2}\sin^{3/2}x\cos x\,dx - \int_{\pi/2}^{\pi}\sin^{3/2}x\cos x\,dx$$

$$= \int_0^{\pi/2} \sin^{3/2} x \, d(\sin x) - \int_{\pi/2}^{\pi} \sin^{3/2} x \, d(\sin x)$$

$$= \left[\frac{2}{5}\sin^{5/2} x\right]_0^{\pi/2} - \left[\frac{2}{5}\sin^{5/2} x\right]_{\pi/2}^{\pi}$$

$$= \frac{2}{5} - \left(-\frac{2}{5}\right) = \frac{4}{5}.$$

Guide $\sqrt{\sin^3 x - \sin^5 x} = \sqrt{\sin^3 x (1-\sin^2 x)}$
$= \sin^{3/2} x |\cos x|.$

Example 4 Compute $\int_0^4 \frac{x+2}{\sqrt{2x+1}} dx.$

Solution

$$\int_0^4 \frac{x+2}{\sqrt{2x+1}} dx \xrightarrow{\text{Let } \sqrt{2x+1}=t} \int_1^3 \frac{\frac{t^2-1}{2}+2}{t} \cdot t \, dt$$

$$= \frac{1}{2} \int_1^3 (t^2+3) \, dt$$

$$= \frac{1}{2} \left[\frac{1}{3}t^3 + 3t\right]_1^3$$

$$= \frac{1}{2}\left[\left(\frac{27}{3}+9\right)-\left(\frac{1}{3}+3\right)\right] = \frac{22}{3}.$$

Guide $x = \frac{t^2-1}{2}$, $dx = t \, dt$. If $x=0$, then $t=1$; if $x=4$, then $t=3$.

Example 5 Prove that if $f(x)$ is an even function and continuous on $[-a, a]$, then

$$\int_{-a}^{a} f(x) dx = 2\int_0^a f(x) dx;$$

if $f(x)$ is an odd function and continuous on $[-a, a]$, then

$$\int_{-a}^{a} f(x) dx = 0.$$

Proof As

$$\int_{-a}^{a} f(x) dx = \int_{-a}^{0} f(x) dx + \int_0^a f(x) dx,$$

and

$$\int_{-a}^{0} f(x) dx \xrightarrow{\text{Let } x=-t} -\int_a^0 f(-t) dt$$

$$= \int_0^a f(-t) dt$$

$$= \int_0^a f(-x) dx,$$

so if $f(x)$ is an even function, then

$$\int_{-a}^{a} f(x)\mathrm{d}x = \int_{0}^{a} f(-x)\mathrm{d}x + \int_{0}^{a} f(x)\mathrm{d}x$$
$$= \int_{0}^{a} [f(-x)+f(x)]\mathrm{d}x$$
$$= \int_{0}^{a} 2f(x)\mathrm{d}x = 2\int_{0}^{a} f(x)\mathrm{d}x;$$

if $f(x)$ is an odd function, $f(-x)+f(x)=0$, then
$$\int_{-a}^{a} f(x)\mathrm{d}x = \int_{0}^{a} [f(-x)+f(x)]\mathrm{d}x = 0.$$

Example 6 Prove that if the function $f(x)$ is continuous on $[0,1]$, then

(1) $\int_{0}^{\pi/2} f(\sin x)\mathrm{d}x = \int_{0}^{\pi/2} f(\cos x)\mathrm{d}x$;

(2) $\int_{0}^{\pi} x f(\sin x)\mathrm{d}x = \dfrac{\pi}{2} \int_{0}^{\pi} f(\sin x)\mathrm{d}x.$

Proof (1) Let $x = \dfrac{\pi}{2} - t$, then
$$\int_{0}^{\pi/2} f(\sin x)\mathrm{d}x = -\int_{\pi/2}^{0} f\left(\sin\left(\dfrac{\pi}{2}-t\right)\right)\mathrm{d}t$$
$$= \int_{0}^{\pi/2} f\left(\sin\left(\dfrac{\pi}{2}-t\right)\right)\mathrm{d}t$$
$$= \int_{0}^{\pi/2} f(\cos x)\mathrm{d}x.$$

(2) Let $x = \pi - t$, then
$$\int_{0}^{\pi} x f(\sin x)\mathrm{d}x = -\int_{\pi}^{0} (\pi-t) f(\sin(\pi-t))\mathrm{d}t$$
$$= \int_{0}^{\pi} (\pi-t) f(\sin(\pi-t))\mathrm{d}t$$
$$= \int_{0}^{\pi} (\pi-t) f(\sin t)\mathrm{d}t$$
$$= \pi \int_{0}^{\pi} f(\sin t)\mathrm{d}t - \int_{0}^{\pi} t f(\sin t)\mathrm{d}t$$
$$= \pi \int_{0}^{\pi} f(\sin x)\mathrm{d}x - \int_{0}^{\pi} x f(\sin x)\mathrm{d}x,$$

so
$$\int_{0}^{\pi} x f(\sin x)\mathrm{d}x = \dfrac{\pi}{2} \int_{0}^{\pi} f(\sin x)\mathrm{d}x.$$

Example 7 If the function
$$f(x) = \begin{cases} x\mathrm{e}^{-x^2}, & x \geqslant 0, \\ \dfrac{1}{1+\cos x}, & -1 < x < 0, \end{cases}$$
then compute $\int_{1}^{4} f(x-2)\mathrm{d}x$.

$$\int_{-a}^{a} f(x)\mathrm{d}x = \int_{0}^{a} f(-x)\mathrm{d}x + \int_{0}^{a} f(x)\mathrm{d}x$$
$$= \int_{0}^{a} [f(-x)+f(x)]\mathrm{d}x$$
$$= \int_{0}^{a} 2f(x)\mathrm{d}x = 2\int_{0}^{a} f(x)\mathrm{d}x;$$

若 $f(x)$ 为奇函数，即 $f(-x)+f(x)=0$，则有
$$\int_{-a}^{a} f(x)\mathrm{d}x = \int_{0}^{a} [f(-x)+f(x)]\mathrm{d}x = 0.$$

例 6 证明：若函数 $f(x)$ 在 $[0,1]$ 上连续，则

(1) $\int_{0}^{\pi/2} f(\sin x)\mathrm{d}x = \int_{0}^{\pi/2} f(\cos x)\mathrm{d}x$；

(2) $\int_{0}^{\pi} x f(\sin x)\mathrm{d}x = \dfrac{\pi}{2} \int_{0}^{\pi} f(\sin x)\mathrm{d}x.$

证明 (1) 令 $x = \dfrac{\pi}{2} - t$，则
$$\int_{0}^{\pi/2} f(\sin x)\mathrm{d}x = -\int_{\pi/2}^{0} f\left(\sin\left(\dfrac{\pi}{2}-t\right)\right)\mathrm{d}t$$
$$= \int_{0}^{\pi/2} f\left(\sin\left(\dfrac{\pi}{2}-t\right)\right)\mathrm{d}t$$
$$= \int_{0}^{\pi/2} f(\cos x)\mathrm{d}x.$$

(2) 令 $x = \pi - t$，则
$$\int_{0}^{\pi} x f(\sin x)\mathrm{d}x = -\int_{\pi}^{0} (\pi-t) f(\sin(\pi-t))\mathrm{d}t$$
$$= \int_{0}^{\pi} (\pi-t) f(\sin(\pi-t))\mathrm{d}t$$
$$= \int_{0}^{\pi} (\pi-t) f(\sin t)\mathrm{d}t$$
$$= \pi \int_{0}^{\pi} f(\sin t)\mathrm{d}t - \int_{0}^{\pi} t f(\sin t)\mathrm{d}t$$
$$= \pi \int_{0}^{\pi} f(\sin x)\mathrm{d}x - \int_{0}^{\pi} x f(\sin x)\mathrm{d}x.$$

所以
$$\int_{0}^{\pi} x f(\sin x)\mathrm{d}x = \dfrac{\pi}{2} \int_{0}^{\pi} f(\sin x)\mathrm{d}x.$$

例 7 设函数
$$f(x) = \begin{cases} x\mathrm{e}^{-x^2}, & x \geqslant 0, \\ \dfrac{1}{1+\cos x}, & -1 < x < 0, \end{cases}$$
计算 $\int_{1}^{4} f(x-2)\mathrm{d}x$.

Solution Let $x-2=t$, then
$$\int_1^4 f(x-2)\,dx = \int_{-1}^2 f(t)\,dt$$
$$= \int_{-1}^0 \frac{1}{1+\cos t}\,dt + \int_0^2 te^{-t^2}\,dt$$
$$= \left[\tan\frac{t}{2}\right]_{-1}^0 - \left[\frac{1}{2}e^{-t^2}\right]_0^2$$
$$= \tan\frac{1}{2} - \frac{1}{2}e^{-4} + \frac{1}{2}.$$

Guide Let $x-2=t$, then $dx=dt$. If $x=1$, then $t=-1$; if $x=4$, then $t=2$.

2. Definite Integration by Parts

If the functions $u(x)$ and $v(x)$ exist continuous derivatives on the interval $[a,b]$, by $(uv)' = u'v + uv'$, then
$$uv' = (uv)' - u'v,$$
so
$$\int_a^b uv'\,dx = [uv]_a^b - \int_a^b u'v\,dx,$$
or
$$\int_a^b u\,dv = [uv]_a^b - \int_a^b v\,du.$$

This is the **formula for definite integration by parts.**

The process of definite integration by parts is as follow:
$$\int_a^b uv'\,dx = \int_a^b u\,dv = [uv]_a^b - \int_a^b v\,du$$
$$= [uv]_a^b - \int_a^b u'v\,dx = \cdots.$$

Example 8 Compute $\int_0^{1/2} \arcsin x\,dx$.

Solution $\int_0^{1/2} \arcsin x\,dx$
$$= [x\arcsin x]_0^{1/2} - \int_0^{1/2} x\,d(\arcsin x)$$
$$= \frac{1}{2}\cdot\frac{\pi}{6} - \int_0^{1/2} \frac{x}{\sqrt{1-x^2}}\,dx$$
$$= \frac{\pi}{12} + \frac{1}{2}\int_0^{1/2} \frac{1}{\sqrt{1-x^2}}\,d(1-x^2)$$
$$= \frac{\pi}{12} + \left[\sqrt{1-x^2}\right]_0^{1/2}$$
$$= \frac{\pi}{12} + \frac{\sqrt{3}}{2} - 1.$$

解 令 $x-2=t$,则
$$\int_1^4 f(x-2)\,dx = \int_{-1}^2 f(t)\,dt$$
$$= \int_{-1}^0 \frac{1}{1+\cos t}\,dt + \int_0^2 te^{-t^2}\,dt$$
$$= \left[\tan\frac{t}{2}\right]_{-1}^0 - \left[\frac{1}{2}e^{-t^2}\right]_0^2$$
$$= \tan\frac{1}{2} - \frac{1}{2}e^{-4} + \frac{1}{2}.$$

提示 令 $x-2=t$,则 $dx=dt$. 当 $x=1$ 时,$t=-1$;当 $x=4$ 时,$t=2$.

2. 定积分的分部积分法

设函数 $u(x),v(x)$ 在区间 $[a,b]$ 上具有连续导数,由 $(uv)' = u'v + uv'$ 得
$$uv' = (uv)' - u'v,$$
于是
$$\int_a^b uv'\,dx = [uv]_a^b - \int_a^b u'v\,dx$$
或
$$\int_a^b u\,dv = [uv]_a^b - \int_a^b v\,du.$$

这就是**定积分的分部积分公式**.

分部积分过程如下:
$$\int_a^b uv'\,dx = \int_a^b u\,dv = [uv]_a^b - \int_a^b v\,du$$
$$= [uv]_a^b - \int_a^b u'v\,dx = \cdots.$$

例8 计算 $\int_0^{1/2} \arcsin x\,dx$.

解 $\int_0^{1/2} \arcsin x\,dx$
$$= [x\arcsin x]_0^{1/2} - \int_0^{1/2} x\,d(\arcsin x)$$
$$= \frac{1}{2}\cdot\frac{\pi}{6} - \int_0^{1/2} \frac{x}{\sqrt{1-x^2}}\,dx$$
$$= \frac{\pi}{12} + \frac{1}{2}\int_0^{1/2} \frac{1}{\sqrt{1-x^2}}\,d(1-x^2)$$
$$= \frac{\pi}{12} + \left[\sqrt{1-x^2}\right]_0^{1/2}$$
$$= \frac{\pi}{12} + \frac{\sqrt{3}}{2} - 1.$$

Example 9 Compute $\int_0^1 e^{\sqrt{x}} dx$.

Solution Let $\sqrt{x}=t$, then
$$\int_0^1 e^{\sqrt{x}} dx = 2\int_0^1 e^t t \, dt = 2\int_0^1 t \, d(e^t)$$
$$= 2[te^t]_0^1 - 2\int_0^1 e^t dt$$
$$= 2e - 2[e^t]_0^1 = 2.$$

Example 10 Compute $\int_0^2 x e^{-2x} dx$.

Solution $\int_0^2 x e^{-2x} dx = \int_0^2 x \, d\left(-\frac{e^{-2x}}{2}\right)$
$$= -\frac{1}{2}[x e^{-2x}]_0^2 + \frac{1}{2}\int_0^2 e^{-2x} dx$$
$$= -\frac{1}{e^4} - \frac{1}{4}\int_0^2 e^{-2x} d(-2x)$$
$$= -\frac{1}{e^4} - \frac{1}{4}[e^{-2x}]_0^2$$
$$= \frac{1}{4} - \frac{5}{4e^4}.$$

Example 11 Compute $\int_1^e x^2 \ln x \, dx$.

Solution $\int_1^e x^2 \ln x \, dx = \int_1^e \ln x \, d\left(\frac{x^3}{3}\right)$
$$= \left[\frac{x^3 \ln x}{3}\right]_1^e - \frac{1}{3}\int_1^e x^3 d(\ln x)$$
$$= \frac{e^3}{3} - \frac{1}{3}\int_1^e x^2 dx$$
$$= \frac{e^3}{3} - \frac{1}{9}[x^3]_1^e$$
$$= \frac{1}{9}(2e^3 + 1).$$

6.4 The Improper Integral

1. The Improper Integral of Infinite Limit

Definition 6.2 Suppose that the function $f(x)$ is continuous on the interval $[a, +\infty)$ and $t > a$. If the limit
$$\lim_{t \to +\infty} \int_a^t f(x) dx$$
exists, then this limit is called the **improper integral** of $f(x)$ on the interval $[a, +\infty)$, signed as $\int_a^{+\infty} f(x) dx$, or

$$\int_a^{+\infty} f(x)\mathrm{d}x = \lim_{t\to+\infty}\int_a^t f(x)\mathrm{d}x.$$

And we say that the improper integral $\int_a^{+\infty} f(x)\mathrm{d}x$ is **convergent**.

If the above-mentioned limit does not exist, the improper integral $\int_a^{+\infty} f(x)\mathrm{d}x$ of the function $f(x)$ on the interval $[a,+\infty)$ is nothing, and in this condition, we say that the improper integral $\int_a^{+\infty} f(x)\mathrm{d}x$ is **non-convergent**.

Similarly, if the function $f(x)$ is continuous on the interval $(-\infty,b]$ and $\lim\limits_{t\to-\infty}\int_t^b f(x)\mathrm{d}x$ exists, then this limit is called the **improper integral** of $f(x)$ on the interval $(-\infty,b]$, signed as $\int_{-\infty}^b f(x)\mathrm{d}x$, or

$$\int_{-\infty}^b f(x)\mathrm{d}x = \lim_{t\to-\infty}\int_t^b f(x)\mathrm{d}x.$$

And we say that the improper integral $\int_{-\infty}^b f(x)\mathrm{d}x$ is **convergent**. If the above-mentioned limit does not exist, we say that the the improper integral $\int_{-\infty}^b f(x)\mathrm{d}x$ is **non-convergent**.

If the function $f(x)$ is continuous on the interval $(-\infty,+\infty)$, the improper integrals $\int_{-\infty}^a f(x)\mathrm{d}x$ and $\int_a^{+\infty} f(x)\mathrm{d}x$ are both convergent, then the sum of the two improper integrals bove is called the **improper integral** of $f(x)$ on the interval $(-\infty,+\infty)$, signed as $\int_{-\infty}^{+\infty} f(x)\mathrm{d}x$, or

$$\int_{-\infty}^{+\infty} f(x)\mathrm{d}x = \int_{-\infty}^a f(x)\mathrm{d}x + \int_a^{+\infty} f(x)\mathrm{d}x.$$

And we say that the improper integral $\int_{-\infty}^{+\infty} f(x)\mathrm{d}x$ is **convergent**.

If any improper integral in the right of the form above is non-convergent, we say that the improper integral $\int_{-\infty}^{+\infty} f(x)\mathrm{d}x$ is **non-convergent**.

$$\int_a^{+\infty} f(x)\mathrm{d}x = \lim_{t\to+\infty}\int_a^t f(x)\mathrm{d}x.$$

这时称反常积分 $\int_a^{+\infty} f(x)\mathrm{d}x$ **收敛**.

如果上述极限不存在，函数 $f(x)$ 在区间 $[a,+\infty)$ 上的反常积分 $\int_a^{+\infty} f(x)\mathrm{d}x$ 就没有意义，此时称反常积分 $\int_a^{+\infty} f(x)\mathrm{d}x$ **发散**.

类似地，如果函数 $f(x)$ 在区间 $(-\infty,b]$ 上连续，且极限 $\lim\limits_{t\to-\infty}\int_t^b f(x)\mathrm{d}x$ $(t<b)$ 存在，则称此极限为 $f(x)$ 在无穷区间 $(-\infty,b]$ 上的**反常积分**，记作 $\int_{-\infty}^b f(x)\mathrm{d}x$，即

$$\int_{-\infty}^b f(x)\mathrm{d}x = \lim_{t\to-\infty}\int_t^b f(x)\mathrm{d}x.$$

这时称反常积分 $\int_{-\infty}^b f(x)\mathrm{d}x$ **收敛**. 如果上述极限不存在，则称反常积分 $\int_{-\infty}^b f(x)\mathrm{d}x$ **发散**.

如果函数 $f(x)$ 在区间 $(-\infty,+\infty)$ 内连续，且反常积分 $\int_{-\infty}^a f(x)\mathrm{d}x$ 和 $\int_a^{+\infty} f(x)\mathrm{d}x$ 都收敛，则称上述两个反常积分的和为 $f(x)$ 在无穷区间 $(-\infty,+\infty)$ 上的**反常积分**，记作 $\int_{-\infty}^{+\infty} f(x)\mathrm{d}x$，即

$$\int_{-\infty}^{+\infty} f(x)\mathrm{d}x = \int_{-\infty}^a f(x)\mathrm{d}x + \int_a^{+\infty} f(x)\mathrm{d}x.$$

这时称反常积分 $\int_{-\infty}^{+\infty} f(x)\mathrm{d}x$ **收敛**.

如果上式右端有一个反常积分发散，则称反常积分 $\int_{-\infty}^{+\infty} f(x)\mathrm{d}x$ **发散**.

Calculation of improper integral: If $F(x)$ is a primitive of $f(x)$, then

$$\int_a^{+\infty} f(x)\,dx = \lim_{t\to+\infty}\int_a^t f(x)\,dx$$
$$= \lim_{t\to+\infty}[F(x)]_a^t$$
$$= \lim_{t\to+\infty}F(t) - F(a)$$
$$= \lim_{x\to+\infty}F(x) - F(a),$$

or simplified as

$$\int_a^{+\infty} f(x)\,dx = [F(x)]_a^{+\infty}$$
$$= \lim_{x\to+\infty}F(x) - F(a).$$

Similarly, we have

$$\int_{-\infty}^b f(x)\,dx = [F(x)]_{-\infty}^b$$
$$= F(b) - \lim_{x\to-\infty}F(x),$$
$$\int_{-\infty}^{+\infty} f(x)\,dx = [F(x)]_{-\infty}^{+\infty}$$
$$= \lim_{x\to+\infty}F(x) - \lim_{x\to-\infty}F(x).$$

Example 1 Compute $\int_{-\infty}^{+\infty}\dfrac{1}{1+x^2}\,dx$.

Solution
$$\int_{-\infty}^{+\infty}\dfrac{1}{1+x^2}\,dx = [\arctan x]_{-\infty}^{+\infty}$$
$$= \lim_{x\to+\infty}\arctan x - \lim_{x\to-\infty}\arctan x$$
$$= \dfrac{\pi}{2} - \left(-\dfrac{\pi}{2}\right) = \pi.$$

Example 2 Compute $\int_0^{+\infty} te^{-pt}\,dt$ (p is a constant and $p>0$).

Solution
$$\int_0^{+\infty} te^{-pt}\,dt = \left[\int te^{-pt}\,dt\right]_0^{+\infty}$$
$$= \left[-\dfrac{1}{p}\int t\,d(e^{-pt})\right]_0^{+\infty}$$
$$= \left[-\dfrac{1}{p}te^{-pt} + \dfrac{1}{p}\int e^{-pt}\,dt\right]_0^{+\infty}$$
$$= \left[-\dfrac{1}{p}te^{-pt} - \dfrac{1}{p^2}e^{-pt}\right]_0^{+\infty}$$
$$= \lim_{t\to+\infty}\left[-\dfrac{1}{p}te^{-pt} - \dfrac{1}{p^2}e^{-pt}\right] + \dfrac{1}{p^2}$$
$$= \dfrac{1}{p^2}.$$

反常积分的计算：如果 $F(x)$ 是 $f(x)$ 的原函数，则

$$\int_a^{+\infty} f(x)\,dx = \lim_{t\to+\infty}\int_a^t f(x)\,dx$$
$$= \lim_{t\to+\infty}[F(x)]_a^t$$
$$= \lim_{t\to+\infty}F(t) - F(a)$$
$$= \lim_{x\to+\infty}F(x) - F(a),$$

或可采用如下简记形式：

$$\int_a^{+\infty} f(x)\,dx = [F(x)]_a^{+\infty}$$
$$= \lim_{x\to+\infty}F(x) - F(a).$$

类似地，有

$$\int_{-\infty}^b f(x)\,dx = [F(x)]_{-\infty}^b$$
$$= F(b) - \lim_{x\to-\infty}F(x),$$
$$\int_{-\infty}^{+\infty} f(x)\,dx = [F(x)]_{-\infty}^{+\infty}$$
$$= \lim_{x\to+\infty}F(x) - \lim_{x\to-\infty}F(x).$$

例 1 计算 $\int_{-\infty}^{+\infty}\dfrac{1}{1+x^2}\,dx$.

解
$$\int_{-\infty}^{+\infty}\dfrac{1}{1+x^2}\,dx = [\arctan x]_{-\infty}^{+\infty}$$
$$= \lim_{x\to+\infty}\arctan x - \lim_{x\to-\infty}\arctan x$$
$$= \dfrac{\pi}{2} - \left(-\dfrac{\pi}{2}\right) = \pi.$$

例 2 计算 $\int_0^{+\infty} te^{-pt}\,dt$（$p$ 是常数且 $p>0$）.

解
$$\int_0^{+\infty} te^{-pt}\,dt = \left[\int te^{-pt}\,dt\right]_0^{+\infty}$$
$$= \left[-\dfrac{1}{p}\int t\,d(e^{-pt})\right]_0^{+\infty}$$
$$= \left[-\dfrac{1}{p}te^{-pt} + \dfrac{1}{p}\int e^{-pt}\,dt\right]_0^{+\infty}$$
$$= \left[-\dfrac{1}{p}te^{-pt} - \dfrac{1}{p^2}e^{-pt}\right]_0^{+\infty}$$
$$= \lim_{t\to+\infty}\left[-\dfrac{1}{p}te^{-pt} - \dfrac{1}{p^2}e^{-pt}\right] + \dfrac{1}{p^2}$$
$$= \dfrac{1}{p^2}.$$

Guide $\lim\limits_{t\to+\infty} te^{-pt} = \lim\limits_{t\to+\infty}\dfrac{t}{e^{pt}} = \lim\limits_{t\to+\infty}\dfrac{1}{pe^{pt}} = 0.$

Example 3 Discuss whether $\int_a^{+\infty} \dfrac{1}{x^p}dx\ (a>0)$ is convergent.

Solution If $p=1$, then
$$\int_a^{+\infty} \dfrac{1}{x^p}dx = \int_a^{+\infty} \dfrac{1}{x}dx = [\ln x]_a^{+\infty} = +\infty.$$

If $p<1$, then
$$\int_a^{+\infty} \dfrac{1}{x^p}dx = \left[\dfrac{1}{1-p}x^{1-p}\right]_a^{+\infty} = +\infty.$$

If $p>1$, then
$$\int_a^{+\infty} \dfrac{1}{x^p}dx = \left[\dfrac{1}{1-p}x^{1-p}\right]_a^{+\infty} = \dfrac{a^{1-p}}{p-1}.$$

So, if $p>1$, the improper integral is convergent to $\dfrac{a^{1-p}}{p-1}$; if $p\leqslant 1$, the improper integral is non-convergent.

2. The Improper Integral of the Unbounded Function

Definition 6.3 If the function $f(x)$ is continuous on interval $(a,b]$, and it's unbounded in right neighborhood of a. If the limit $\lim\limits_{t\to a^+}\int_t^b f(x)dx$ exists, then the limit is called the **improper integral** of the function $f(x)$ on $(a,b]$, signed as $\int_a^b f(x)dx$ also, or
$$\int_a^b f(x)dx = \lim\limits_{t\to a^+}\int_t^b f(x)dx.$$

And we say that the improper integral $\int_a^b f(x)dx$ is **convergent.**

If the above-mentioned limit does not exist, we say that the improper integral $\int_a^b f(x)dx$ is **non-convergent.**

Similarly, if function $f(x)$ is continuous on the interval $[a,b)$, and it's unbounded in left neighborhood of b. If the limit $\lim\limits_{t\to b^-}\int_a^t f(x)dx$ exists, then the limit is called the **improper integral** of the function $f(x)$ on $[a,b)$, signed as $\int_a^b f(x)dx$ also, or

提示 $\lim\limits_{t\to+\infty} te^{-pt} = \lim\limits_{t\to+\infty}\dfrac{t}{e^{pt}} = \lim\limits_{t\to+\infty}\dfrac{1}{pe^{pt}} = 0.$

例3 讨论 $\int_a^{+\infty} \dfrac{1}{x^p}dx\ (a>0)$ 的敛散性.

解 若 $p=1$，则有
$$\int_a^{+\infty} \dfrac{1}{x^p}dx = \int_a^{+\infty} \dfrac{1}{x}dx = [\ln x]_a^{+\infty} = +\infty;$$

若 $p<1$，则有
$$\int_a^{+\infty} \dfrac{1}{x^p}dx = \left[\dfrac{1}{1-p}x^{1-p}\right]_a^{+\infty} = +\infty;$$

若 $p>1$，则有
$$\int_a^{+\infty} \dfrac{1}{x^p}dx = \left[\dfrac{1}{1-p}x^{1-p}\right]_a^{+\infty} = \dfrac{a^{1-p}}{p-1}.$$

因此，若 $p>1$，则此反常积分收敛，其值为 $\dfrac{a^{1-p}}{p-1}$；若 $p\leqslant 1$，则此反常积分发散.

2. 无界函数的反常积分

定义6.3 设函数 $f(x)$ 在区间 $(a,b]$ 上连续，而在点 a 的右侧附近无界. 如果极限 $\lim\limits_{t\to a^+}\int_t^b f(x)dx$ 存在，则称此极限为函数 $f(x)$ 在 $(a,b]$ 上的**反常积分**，仍然记作 $\int_a^b f(x)dx$，即
$$\int_a^b f(x)dx = \lim\limits_{t\to a^+}\int_t^b f(x)dx.$$

这时称反常积分 $\int_a^b f(x)dx$ **收敛**.

如果上述极限不存在，就称反常积分 $\int_a^b f(x)dx$ **发散**.

类似地，设函数 $f(x)$ 在区间 $[a,b)$ 上连续，而在点 b 的左侧附近无界. 如果极限 $\lim\limits_{t\to b^-}\int_a^t f(x)dx$ 存在，则称此极限为函数 $f(x)$ 在 $[a,b)$ 上的**反常积分**，仍然记作 $\int_a^b f(x)dx$，即

$$\int_a^b f(x)\,dx = \lim_{t \to b^-} \int_a^t f(x)\,dx.$$

And we say that the improper integral $\int_a^b f(x)\,dx$ is **convergent**.

If the above-mentioned limit does not exist, we say that the improper integral $\int_a^b f(x)\,dx$ is **non-convergent**.

If the function $f(x)$ is continuous on the interval $[a,b]$ except the point c ($a<c<b$), and it's unbounded in neighborhood of c. If the improper integrals $\int_a^c f(x)\,dx$ and $\int_c^b f(x)\,dx$ are both convergent, then define

$$\int_a^b f(x)\,dx = \int_a^c f(x)\,dx + \int_c^b f(x)\,dx,$$

else, we say that the improper integral $\int_a^b f(x)\,dx$ is **non-convergent**.

If the function $f(x)$ is unbounded in any neighborhood of a, then a is called a **singular point** or **unbounded point** of $f(x)$.

Calculation of improper integral: If $F(x)$ is a primitive of $f(x)$, when a is a singular point, then

$$\begin{aligned}\int_a^b f(x)\,dx &= \lim_{t \to a^+} \int_t^b f(x)\,dx \\ &= \lim_{t \to a^+} [F(x)]_t^b \\ &= F(b) - \lim_{t \to a^+} F(t) \\ &= F(b) - \lim_{x \to a^+} F(x),\end{aligned}$$

or simplified as

$$\begin{aligned}\int_a^b f(x)\,dx &= [F(x)]_a^b \\ &= F(b) - \lim_{x \to a^+} F(x).\end{aligned}$$

Similarly, when b is a singular point, then

$$\begin{aligned}\int_a^b f(x)\,dx &= [F(x)]_a^b \\ &= \lim_{x \to b^-} F(x) - F(a);\end{aligned}$$

when c ($a<c<b$) is a singular point, then

$$\begin{aligned}\int_a^b f(x)\,dx &= \int_a^c f(x)\,dx + \int_c^b f(x)\,dx \\ &= [\lim_{x \to c^-} F(x) - F(a)] \\ &\quad + [F(b) - \lim_{x \to c^+} F(x)].\end{aligned}$$

$$\int_a^b f(x)\,dx = \lim_{t \to b^-} \int_a^t f(x)\,dx.$$

这时称反常积分 $\int_a^b f(x)\,dx$ **收敛**.

如果上述极限不存在,就称反常积分 $\int_a^b f(x)\,dx$ **发散**.

设函数 $f(x)$ 在区间 $[a,b]$ 上除点 c ($a<c<b$) 外连续,而在点 c 的附近无界. 如果两个反常积分 $\int_a^c f(x)\,dx$ 与 $\int_c^b f(x)\,dx$ 都收敛,则定义

$$\int_a^b f(x)\,dx = \int_a^c f(x)\,dx + \int_c^b f(x)\,dx;$$

否则,就称反常积分 $\int_a^b f(x)\,dx$ **发散**.

如果函数 $f(x)$ 在点 a 的附近无界,那么点 a 称为函数 $f(x)$ 的**瑕点**,也称为**无界点**.

反常积分的计算:如果 $F(x)$ 为 $f(x)$ 的原函数,当 a 为瑕点时,有

$$\begin{aligned}\int_a^b f(x)\,dx &= \lim_{t \to a^+} \int_t^b f(x)\,dx \\ &= \lim_{t \to a^+} [F(x)]_t^b \\ &= F(b) - \lim_{t \to a^+} F(t) \\ &= F(b) - \lim_{x \to a^+} F(x),\end{aligned}$$

或采用如下简记形式:

$$\begin{aligned}\int_a^b f(x)\,dx &= [F(x)]_a^b \\ &= F(b) - \lim_{x \to a^+} F(x).\end{aligned}$$

类似地,当 b 为瑕点时,有

$$\begin{aligned}\int_a^b f(x)\,dx &= [F(x)]_a^b \\ &= \lim_{x \to b^-} F(x) - F(a);\end{aligned}$$

当 c ($a<c<b$) 为瑕点时,有

$$\begin{aligned}\int_a^b f(x)\,dx &= \int_a^c f(x)\,dx + \int_c^b f(x)\,dx \\ &= [\lim_{x \to c^-} F(x) - F(a)] \\ &\quad + [F(b) - \lim_{x \to c^+} F(x)].\end{aligned}$$

Example 4 Compute $\int_0^a \frac{1}{\sqrt{a^2-x^2}}dx$.

Solution As $\lim\limits_{x\to a^-}\frac{1}{\sqrt{a^2-x^2}}=+\infty$, so a is a singular point of $\frac{1}{\sqrt{a^2-x^2}}$, then

$$\int_0^a \frac{1}{\sqrt{a^2-x^2}}dx = \left[\arcsin\frac{x}{a}\right]_0^a$$
$$= \lim_{x\to a^-}\arcsin\frac{x}{a} - 0 = \frac{\pi}{2}.$$

Example 5 Discuss whether $\int_{-1}^1 \frac{1}{x^2}dx$ is convergent.

Solution The function $\frac{1}{x^2}$ is continuous on $[-1,1]$ except $x=0$, and $\lim\limits_{x\to 0}\frac{1}{x^2}=\infty$.

As

$$\int_{-1}^0 \frac{1}{x^2}dx = \left[-\frac{1}{x}\right]_{-1}^0$$
$$= \lim_{x\to 0^-}\left(-\frac{1}{x}\right) - 1 = +\infty,$$

or $\int_{-1}^0 \frac{1}{x^2}dx$ is non-convergent, then $\int_{-1}^1 \frac{1}{x^2}dx$ is non-convergent.

Example 6 Discuss whether $\int_a^b \frac{dx}{(x-a)^q}$ is convergent.

Solution If $q=1$, then

$$\int_a^b \frac{dx}{(x-a)^q} = \int_a^b \frac{dx}{x-a}$$
$$= [\ln(x-a)]_a^b = +\infty.$$

If $q>1$, then

$$\int_a^b \frac{dx}{(x-a)^q} = \left[\frac{1}{1-q}(x-a)^{1-q}\right]_a^b = +\infty.$$

If $q<1$, then

$$\int_a^b \frac{dx}{(x-a)^q} = \left[\frac{1}{1-q}(x-a)^{1-q}\right]_a^b$$
$$= \frac{1}{1-q}(b-a)^{1-q}.$$

So, if $q<1$, the improper integral is convergent to $\frac{1}{1-q}(b-a)^{1-q}$; if $q\geq 1$, the improper integral is non-convergent.

例 4 计算 $\int_0^a \frac{1}{\sqrt{a^2-x^2}}dx$.

解 因为 $\lim\limits_{x\to a^-}\frac{1}{\sqrt{a^2-x^2}}=+\infty$,所以点 a 为 $\frac{1}{\sqrt{a^2-x^2}}$ 的瑕点. 于是

$$\int_0^a \frac{1}{\sqrt{a^2-x^2}}dx = \left[\arcsin\frac{x}{a}\right]_0^a$$
$$= \lim_{x\to a^-}\arcsin\frac{x}{a} - 0 = \frac{\pi}{2}.$$

例 5 讨论 $\int_{-1}^1 \frac{1}{x^2}dx$ 的敛散性.

解 函数 $\frac{1}{x^2}$ 在区间$[-1,1]$上除点 $x=0$ 外连续,且 $\lim\limits_{x\to 0}\frac{1}{x^2}=\infty$.

由于

$$\int_{-1}^0 \frac{1}{x^2}dx = \left[-\frac{1}{x}\right]_{-1}^0$$
$$= \lim_{x\to 0^-}\left(-\frac{1}{x}\right) - 1 = +\infty,$$

即 $\int_{-1}^0 \frac{1}{x^2}dx$ 发散,所以 $\int_{-1}^1 \frac{1}{x^2}dx$ 发散.

例 6 讨论 $\int_a^b \frac{dx}{(x-a)^q}$ 的敛散性.

解 若 $q=1$,则有

$$\int_a^b \frac{dx}{(x-a)^q} = \int_a^b \frac{dx}{x-a}$$
$$= [\ln(x-a)]_a^b = +\infty;$$

若 $q>1$,则有

$$\int_a^b \frac{dx}{(x-a)^q} = \left[\frac{1}{1-q}(x-a)^{1-q}\right]_a^b = +\infty;$$

若 $q<1$,则有

$$\int_a^b \frac{dx}{(x-a)^q} = \left[\frac{1}{1-q}(x-a)^{1-q}\right]_a^b$$
$$= \frac{1}{1-q}(b-a)^{1-q}.$$

因此,当 $q<1$ 时,此反常积分收敛,其值为 $\frac{1}{1-q}(b-a)^{1-q}$;当 $q\geq 1$ 时,此反常积分发散.

6.5 Applications of the Definite Integral
6.5 定积分的应用

The definite integral comes from solving practical problems. In the process of the definite integral, we studied its geometric and physical significance, found that the mathematical model of the definite integral is derived from "the limit of the sum". After mastering the method of definite integral operation, we will apply the computation method of integral calculus to practical problems.

1. The Infinitesimal Method

By recalling the calculation process for the area of the trapezoid with curve, we can better understand the definition of definite integral. Thereafter, we can summarize the basic idea in applying the definite integral to solve practical problems.

Given the function $y=f(x)\geqslant 0(x\in[a,b])$ is continuous, then throughout the geometric meaning of the definite integral we know that

$$S=\int_a^b f(x)\mathrm{d}x$$

is the area of trapezoid with curve based $[a,b]$, where $y=f(x)$ is the curve side. The calculation process is summarized as follows:

Divide the interval $[a,b]$ into n sub-intervals $[x_{i-1},x_i]$ $(i=1,2,\cdots,n)$, then the trapezoid with curve is divided into n small trapezoids with curve. With a width of the small rectangle instead of the ditty trapezoid with curve, then, Δx_i as its base and $f(\xi_i)$ as its height and the area of the small rectangle is

$$\Delta S_i=f(\xi_i)\Delta x_i.$$

So the approximation of S is

$$S\approx \sum_{i=1}^n f(\xi_i)\Delta x_i.$$

Let $n\to\infty$, take the limit, then

$$S=\lim_{\lambda\to 0}\sum_{i=1}^n f(\xi_i)\Delta x_i=\int_a^b f(x)\mathrm{d}x.$$

1. 微元法

回忆曲边梯形面积的计算过程，体会定积分的定义，由此总结得出应用定积分解决实际问题的基本思想．

设 $y=f(x)\geqslant 0(x\in[a,b])$ 为连续函数，那么由定积分的几何意义可知

$$S=\int_a^b f(x)\mathrm{d}x$$

是以 $[a,b]$ 为底，$y=f(x)$ 为曲边的曲边梯形的面积．计算过程归纳如下：

将区间 $[a,b]$ 分割为 n 个小区间 $[x_{i-1},x_i]$ $(i=1,2,\cdots,n)$，则曲边梯形被划分为 n 个小曲边梯形．用等宽的小矩形代替小曲边梯形，而小矩形以 Δx_i 为底，$f(\xi_i)$ 为高，它的面积为

$$\Delta S_i=f(\xi_i)\Delta x_i.$$

于是 S 的近似值为

$$S\approx \sum_{i=1}^n f(\xi_i)\Delta x_i.$$

令 $n\to\infty$，取极限得

$$S=\lim_{\lambda\to 0}\sum_{i=1}^n f(\xi_i)\Delta x_i=\int_a^b f(x)\mathrm{d}x.$$

In the process, the key is:

(1) Use the differential $dS = f(x)dx$ to represent the approximation of the area of trapezoid with curve on the sub-interval $[x, x+dx]$: $\Delta S \approx f(x)dx$. We say that $dS = f(x)dx$ is the **area element** of the trapezoid with curve.

(2) The area S of trapezoid with curve based $[a,b]$ is the definite integral with $dS = f(x)dx$ as the integrand expression, and $[a,b]$ as the integral interval, that is

$$S = \int_a^b f(x)dx.$$

In general, to find a certain amount U, first this quantity will be distributed in a certain interval $[a,b]$; then take a small interval $[x, x+dx]$ on the interval $[a,b]$, and find the approximate value of the distribution on this interval, then we get the microelement of U, dU, given $dU = u(x)dx$; finally, take $u(x)dx$ as the integrand expression and $[a,b]$ as the integral interval, then find the definite integral:

$$U = \int_a^b f(x)dx.$$

The method of finding a quantity in this method is called the **infinitesimal method**.

2. Applications in Geometry

1) The Area of Plane Figure

● Rectangular Situation

Given the plane figure is surrounded by two curves $y = f_{up}(x)$ and $y = f_{down}(x)$, and two lines $x = a$ and $x = b$ (Figure 6.5), then the area element is

$$[f_{up}(x) - f_{down}(x)]dx,$$

so the area of the plane figure is

$$S = \int_a^b [f_{up}(x) - f_{down}(x)]dx. \quad (6.1)$$

Similarly, suppose that plane figure is surrounded by two curves $x = \varphi_{left}(y)$ and $x = \varphi_{right}(y)$, and two lines $y = d$ and $y = c$ (Figure 6.6), then the area of the plane figure is

$$S = \int_c^d [\varphi_{right}(y) - \varphi_{left}(y)]dy. \quad (6.2)$$

上述过程的关键在于：

（1）用微分 $dS = f(x)dx$ 表示小区间 $[x, x+dx]$ 上小曲边梯形面积的近似值：$\Delta S \approx f(x)dx$. 称 $dS = f(x)dx$ 为曲边梯形的**面积微元**.

（2）以$[a,b]$为底的曲边梯形的面积 S 就是以面积微元 $dS = f(x)dx$ 为被积表达式，以$[a,b]$为积分区间的定积分，即

$$S = \int_a^b f(x)dx.$$

一般情况下，为了求某一量 U，先将此量分布在某一区间$[a,b]$上；然后，在$[a,b]$上取一小区间$[x, x+dx]$，求分布在这一小区间上的量的近似值，得到 U 的微元 $dU(x)$，设 $dU(x) = u(x)dx$；最后，以 $u(x)dx$ 为被积表达式，以$[a,b]$为积分区间求定积分，即得

$$U = \int_a^b f(x)dx.$$

用这一方法求某一量的值的方法称为**微元法**.

2. 在几何中的应用

1）平面图形的面积

● 直角坐标情形

设平面图形由两条曲线 $y = f_{up}(x)$ 与 $y = f_{down}(x)$ 及两条直线 $x = a$ 与 $x = b$ 所围成（图 6.5），则面积微元为

$$[f_{up}(x) - f_{down}(x)]dx.$$

于是平面图形的面积为

$$S = \int_a^b [f_{up}(x) - f_{down}(x)]dx. \quad (6.1)$$

类似地，由两条曲线 $x = \varphi_{left}(y)$ 与 $x = \varphi_{right}(y)$ 及两条直线 $y = d$ 与 $y = c$ 所围成的平面图形（图 6.6），其面积为

$$S = \int_c^d [\varphi_{right}(y) - \varphi_{left}(y)]dy. \quad (6.2)$$

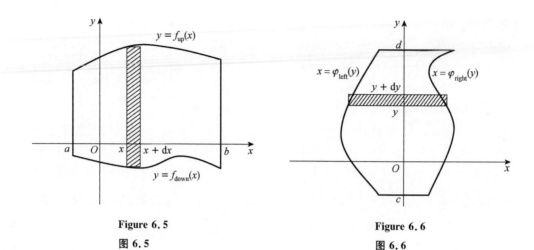

Figure 6.5
图 6.5

Figure 6.6
图 6.6

Example 1 Find the area of the region enclosed by the parabola $y=x^2$ and the line $y=x$.

Solution The Figure 6.7 shows the shape of the parabola and the line.

例 1 求由抛物线 $y=x^2$ 与直线 $y=x$ 所围成的平面图形的面积.

解 抛物线 $y=x^2$ 与直线 $y=x$ 的图像如图 6.7 所示.

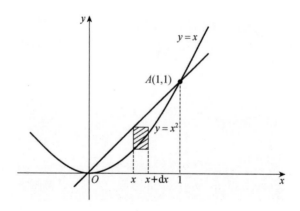

Figure 6.7
图 6.7

First, let's figure out the intersection point of these two lines. We solve the equations $\begin{cases} y=x^2, \\ y=x, \end{cases}$ and get two solutions: $x=0, y=0$ and $x=1, y=1$, then the intersection points are $(0,0)$, $(1,1)$.

Take the abscissa x as the integral variable, its variable interval is $[0,1]$, and $f_{up}(x)=x$, $f_{down}(x)=x^2$. From formula (6.1), we get the area:

先求出这两条线的交点. 我们解方程组 $\begin{cases} y=x^2, \\ y=x, \end{cases}$ 得两个解: $x=0, y=0$ 及 $x=1, y=1$, 即交点为 $(0,0)(1,1)$.

取横坐标 x 为积分变量, 它的变化区间为 $[0,1]$, 且 $f_{up}(x)=x$, $f_{down}(x)=x^2$, 由公式(6.1)得所求面积为

$$S = \int_0^1 (x-x^2)\,dx$$
$$= \left[\frac{1}{2}x^2 - \frac{1}{3}x^3\right]_0^1$$
$$= \frac{1}{2} - \frac{1}{3} = \frac{1}{6}.$$

Example 2 Find the area of the region enclosed by the parabola $y^2 = x$, and the line $y = x - 2$.

Solution The Figure 6.8 shows the shape of the parabola $y^2 = x$ and the line $y = x - 2$.

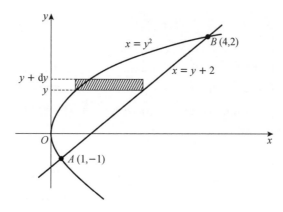

Figure 6.8

First, let's figure out the intersection point of these two lines. We solve the equations $\begin{cases} y^2 = x, \\ y = x - 2, \end{cases}$ and get two solutions: $x = 1, y = -1$ and $x = 4, y = 2$, then the intersection points are $(1, -1)$ and $(4, 2)$.

Method 1 Take the ordinate y as the integral variable, its variable interval is $[-1, 2]$, and $\varphi_{\text{left}}(y) = y^2$, $\varphi_{\text{right}}(y) = y + 2$. From formula (6.2), we get the area:

$$S = \int_{-1}^{2} (y + 2 - y^2)\,dy$$
$$= \left[\frac{1}{2}y^2 + 2y - \frac{1}{3}y^3\right]_{-1}^{2} = \frac{9}{2}.$$

Method 2 If take the abscissa x as the integral variable, and its variable interval is $[0, 4]$. We draw a line passing through the point A, which is perpendicular to x-axis, then we find that the lower curves on the left and

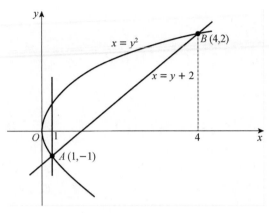

Figure 6.9

图 6.9

right are not the same (Figure 6.9). According to the interval of definite integral is additive, we divide $[0,4]$ into $[0,1]$ and $[1,4]$, and divide the original figure into the left and right sides of the graph, and compute the area respectively.

For the left figure, $f_{\text{up}}(x)=\sqrt{x}$, $f_{\text{down}}(x)=-\sqrt{x}$, so the area of the left figure is

$$S_{\text{left}} = \int_0^1 [\sqrt{x}-(-\sqrt{x})]\mathrm{d}x$$
$$= \int_0^1 2\sqrt{x}\,\mathrm{d}x = 2\left[\frac{2}{3}x^{\frac{3}{2}}\right]_0^1 = \frac{4}{3};$$

For the right figure, $f_{\text{up}}(x)=\sqrt{x}$, $f_{\text{down}}(x)=x-2$, so the area of the right figure is

$$S_{\text{right}} = \int_1^4 (\sqrt{x}-x+2)\,\mathrm{d}x$$
$$= \left[\frac{2}{3}x^{\frac{3}{2}}-\frac{1}{2}x^2+2x\right]_1^4 = \frac{19}{6}.$$

So the area is

$$S = S_{\text{left}} + S_{\text{right}} = \frac{4}{3}+\frac{19}{6} = \frac{9}{2}.$$

- **Polar Situation**

We know the formula of the sectorial area is

$$S = \frac{1}{2}R^2\theta,$$

where R is the radius of the circle, θ is the central angle. The graph rounded by the curve $\rho=\varphi(\theta)$ and the ray $\theta=\alpha, \theta=\beta$ is called a **sector with curve**. Because you can't

可加性,将 $[0,4]$ 划分为 $[0,1][1,4]$ 两个区间,对应的将原图形分为左、右两个图形,分别求其面积.

对于左边图形, $f_{\text{up}}(x)=\sqrt{x}$, $f_{\text{down}}(x)=-\sqrt{x}$, 所以左边图形的面积为

$$S_{\text{left}} = \int_0^1 [\sqrt{x}-(-\sqrt{x})]\mathrm{d}x$$
$$= \int_0^1 2\sqrt{x}\,\mathrm{d}x = 2\left[\frac{2}{3}x^{\frac{3}{2}}\right]_0^1 = \frac{4}{3};$$

对于右边图形, $f_{\text{up}}(x)=\sqrt{x}$, $f_{\text{down}}(x)=x-2$, 所以右边图形的面积为

$$S_{\text{right}} = \int_1^4 (\sqrt{x}-x+2)\,\mathrm{d}x$$
$$= \left[\frac{2}{3}x^{\frac{3}{2}}-\frac{1}{2}x^2+2x\right]_1^4 = \frac{19}{6}.$$

故所求的面积为

$$S = S_{\text{left}} + S_{\text{right}} = \frac{4}{3}+\frac{19}{6} = \frac{9}{2}.$$

- **极坐标情形**

我们知道扇形面积公式为

$$S = \frac{1}{2}R^2\theta,$$

其中 R 为半径, θ 为圆心角. 由曲线 $\rho=\varphi(\theta)$ 及射线 $\theta=\alpha, \theta=\beta$ 所围成的平面图形称为**曲边扇形**. 因为不能保证 $\rho=\varphi(\theta)$ 为圆弧,所以

guarantee that $\rho = \varphi(\theta)$ is the arc, you can't apply the formula of the sectorial area directly.

Applying the infinitesimal method, we divide the sector with curve into a number of small ones through using the rays staring from the pole. As the small sectors with curve are small enough, they can be regarded as small sectors approximatively. Then, we can write the area element of the sector with curve:
$$dS = \frac{1}{2}[\varphi(\theta)]^2 d\theta.$$

So the area is
$$S = \int_\alpha^\beta \frac{1}{2}[\varphi(\theta)]^2 d\theta. \tag{6.3}$$

Example 3 Find the area of the region bounded by three-leaf rose line $r = a\sin 3\theta$.

Solution The Figure 6.10 shows a leaf of the three-leaf rose line, and the area of the region bounded by the three-leaf rose line is three times the area of the pattern, so the area is
$$S = 3\int_{-\pi/6}^{\pi/6} \frac{1}{2}(a\sin 3\theta)^2 d\theta$$
$$= 3\int_0^{\pi/6} a^2 \sin^2 3\theta d\theta$$
$$= a^2 \int_0^{\pi/6} \sin^2 3\theta d\theta.$$

Let $3\theta = t$, then when $\theta = 0, t = 0$; when $\theta = \frac{\pi}{6}, t = \frac{\pi}{2}$. So the area is
$$S = a^2 \int_0^{\pi/2} \sin^2 t dt$$
$$= a^2 \frac{x}{2} \cdot \frac{1}{2} = \frac{\pi a^2}{4}.$$

不能直接应用扇形面积公式来求曲边扇形的面积.

应用微元法,由极点引出射线将曲边扇形进行分割,得到若干小曲边扇形.当切割得足够小时,将每个小曲边扇形近似看作小扇形.于是可以写出曲边扇形的面积微元为
$$dS = \frac{1}{2}[\varphi(\theta)]^2 d\theta,$$

从而曲边扇形的面积为
$$S = \int_\alpha^\beta \frac{1}{2}[\varphi(\theta)]^2 d\theta. \tag{6.3}$$

例3 求三叶玫瑰线 $r = a\sin 3\theta$ 所围平面图形的面积.

解 图 6.10 给出了三叶玫瑰线的一叶,三叶玫瑰线所围平面图形的面积为所示图形面积的3倍,所以所求的面积为
$$S = 3\int_{-\pi/6}^{\pi/6} \frac{1}{2}(a\sin 3\theta)^2 d\theta$$
$$= 3\int_0^{\pi/6} a^2 \sin^2 3\theta d\theta$$
$$= a^2 \int_0^{\pi/6} \sin^2 3\theta d\theta.$$

令 $3\theta = t$,则当 $\theta = 0$ 时,$t = 0$;当 $\theta = \frac{\pi}{6}$ 时,$t = \frac{\pi}{2}$. 所以所求的面积为
$$S = a^2 \int_0^{\pi/2} \sin^2 t dt$$
$$= a^2 \frac{x}{2} \cdot \frac{1}{2} = \frac{\pi a^2}{4}.$$

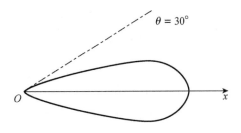

Figure 6.10
图 6.10

Example 4 Find the area of the region rounded by the lemniscate $r^2 = a^2\cos2\theta$.

Solution The Figure 6.11 shows the lemniscate, which is twice the size of the figure on the right, so the area is

$$\begin{aligned}S &= 2\int_{-\pi/4}^{\pi/4}\frac{1}{2}a^2\cos2\theta\mathrm{d}\theta\\ &= 2a^2\int_0^{\pi/4}\cos2\theta\mathrm{d}\theta\\ &= a^2\int_0^{\pi/4}\cos2\theta\mathrm{d}(2\theta)\\ &= a^2\left[\sin2\theta\right]_0^{\pi/4} = a^2.\end{aligned}$$

例 4 计算双纽线 $r^2 = a^2\cos2\theta$ 所围平面图形的面积.

解 图 6.11 给出了双纽线, 双纽线所围平面图形的面积为右边图形面积的 2 倍, 所以所求的面积为

$$\begin{aligned}S &= 2\int_{-\pi/4}^{\pi/4}\frac{1}{2}a^2\cos2\theta\mathrm{d}\theta\\ &= 2a^2\int_0^{\pi/4}\cos2\theta\mathrm{d}\theta\\ &= a^2\int_0^{\pi/4}\cos2\theta\mathrm{d}(2\theta)\\ &= a^2\left[\sin2\theta\right]_0^{\pi/4} = a^2.\end{aligned}$$

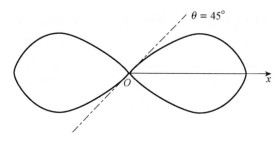

Figure 6.11
图 6.11

2) The Volume of a Solid

Although the geometrical significance of the definite integral is the area of trapezoid with curved, we can also find the volume of some special solid with definite integral. The particularity of these special solid lies in that their volume element can be expressed explicitly.

● The Volume of a Solid of Revolution

A **solid of revolution** is a solid figure obtained by rotating a plane curve around a straight line that lies on the same plane. The straight line is called the **axis of revolution**.

In general, the solids of revolution include columns, cones, round tables, and spheres.

A solid of revolution can be regarded as a solid which obtained by rotating about x-axis the region bounded by the curve $y = f(x)$, the lines $x = a$, $x = b$ and x-axis (Figure 6.12).

2) 立体的体积

虽然定积分的几何意义为曲边梯形的面积, 但是应用定积分也可求出一些特殊立体的体积, 这些特殊立体的特点是能写出其体积微元.

● 旋转体的体积

旋转体就是由一个平面图形绕所在平面内一条直线旋转一周而形成的立体, 这条直线叫作**旋转轴**.

常见的旋转体有圆柱、圆锥、圆台、球体.

旋转体可以看作由连续曲线 $y = f(x)$, 直线 $x = a$, $x = b$ 及 x 轴所围成的曲边梯形绕 x 轴旋转一周而形成的立体(图 6.12).

6.5 Applications of the Definite Integral
6.5 定积分的应用

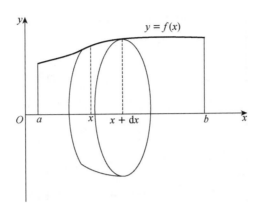

Figure 6.12
图 6.12

If we take an arbitrary sub-internal $[x, x+\mathrm{d}x]$ on the interval $[a,b]$, the number of the sectional volume of the solid of revolution corresponding to the sub-interval $[x, x+\mathrm{d}x]$ approximates to $\Delta V = \pi [f(x)]^2 \mathrm{d}x$, so the volume element is

$$\mathrm{d}V = \pi [f(x)]^2 \mathrm{d}x.$$

Thus, the volume of the solid of revolution is

$$V = \int_a^b \pi [f(x)]^2 \mathrm{d}x. \tag{6.4}$$

Example 5 A right triangle is surrounded by the line passing through the origin O and $P(a,b)$, the line $x=a$ and x-axis. We rotate it around x-axis to get a cone with the bottom radius b and height a (Figure 6.13). Find the volume of the cone.

在区间 $[a,b]$ 上任取小区间 $[x, x+\mathrm{d}x]$，该小区间对应的部分旋转体体积的近似值为 $\Delta V = \pi [f(x)]^2 \mathrm{d}x$，于是体积微元为

$$\mathrm{d}V = \pi [f(x)]^2 \mathrm{d}x.$$

所以旋转体的体积为

$$V = \int_a^b \pi [f(x)]^2 \mathrm{d}x. \tag{6.4}$$

例 5 连接坐标原点 O 及点 $P(a,b)$ 的直线、直线 $x=a$ 及 x 轴围成一个直角三角形．将它绕 x 轴旋转得到一个底半径为 b，高为 a 的圆锥体（图 6.13）．计算这个圆锥体的体积．

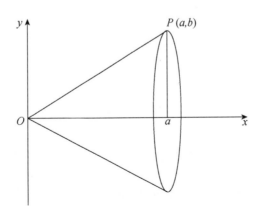

Figure 6.13
图 6.13

Solution The equation of the line at the hypotenuse of the right triangle is
$$y = \frac{b}{a}x.$$
The volume of the cone is
$$V = \int_0^a \pi \left(\frac{b}{a}x\right)^2 dx$$
$$= \frac{\pi b^2}{a^2}\left[\frac{1}{3}x^3\right]_0^a = \frac{1}{3}\pi ab^2.$$

Example 6 Find the volume of the solid of revolution that is obtained by rotating about x-axis the region bounded by the curve $y = \cos x \left(0 \leqslant x \leqslant \frac{\pi}{2}\right)$ and x-axis (Figure 6.14).

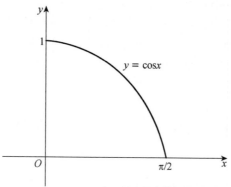

Figure 6.14

Solution The volume of the solid of revolution is
$$V = \int_0^{\pi/2} \pi \cos^2 x \, dx$$
$$= \pi \int_0^{\pi/2} \frac{1+\cos 2x}{2} dx$$
$$= \frac{\pi}{2} \int_0^{\pi/2} (1+\cos 2x) dx = \frac{\pi^2}{4}.$$

- **The Volume of a Solid with Known Paralled Section Area**

Suppose that the projection interval of a solid is $[a,b]$ on x-axis. We know the area of the section that is perpendicular to x-axis and passes through the point x, and write it as $S(x)$. Then, the volume element is $S(x)dx$, and the volume of the solid is

$$V = \int_a^b S(x)\,\mathrm{d}x. \qquad (6.5)$$

Example 7 A plane passes through the bottom circle center of a cylinder with radius R, and the interfacial angle between the plane and the bottom surface is α. Find the volume of the solid obtained by truncating the cylinder in the plane.

Solution If we take the intersecting line between the plane and the bottom surface of the cylinder as x-axis, the line that passes through the center of the circle and is perpendicular to x-axis in bottom surface as y-axis, then the equation of bottom circle is $x^2 + y^2 = R^2$ (Figure 6.15). The cross section of the solid, which passes through x and is perpendicular to x-axis, is a right triangle. And the length of two right-angle sides are $\sqrt{R^2 - x^2}$ and $\sqrt{R^2 - x^2}\tan\alpha$ respectively, then the area of the section is

$$S(x) = \frac{1}{2}(R^2 - x^2)\tan\alpha.$$

Thus, the volume of the solid is

$$\begin{aligned} V &= \int_{-R}^{R} \frac{1}{2}(R^2 - x^2)\tan\alpha\,\mathrm{d}x \\ &= \frac{1}{2}\tan\alpha\left[R^2 x - \frac{1}{3}x^3\right]_{-R}^{R} \\ &= \frac{2}{3}R^3\tan\alpha. \end{aligned}$$

$$V = \int_a^b S(x)\,\mathrm{d}x. \qquad (6.5)$$

例 7 一平面经过半径为 R 的圆柱体的底圆圆心,并与底面交成角 α. 计算这个平面截圆柱所得立体的体积.

解 如果取该平面与圆柱体的底面的交线为 x 轴,底面上过圆心且垂直于 x 轴的直线为 y 轴,那么底圆的方程为 $x^2 + y^2 = R^2$(图 6.15). 该立体的过点 x 且垂直于 x 轴的截面是一个直角三角形,两个直角边分别为 $\sqrt{R^2 - x^2}$ 及 $\sqrt{R^2 - x^2}\tan\alpha$,从而截面面积为

$$S(x) = \frac{1}{2}(R^2 - x^2)\tan\alpha.$$

于是所求的立体体积为

$$\begin{aligned} V &= \int_{-R}^{R} \frac{1}{2}(R^2 - x^2)\tan\alpha\,\mathrm{d}x \\ &= \frac{1}{2}\tan\alpha\left[R^2 x - \frac{1}{3}x^3\right]_{-R}^{R} \\ &= \frac{2}{3}R^3\tan\alpha. \end{aligned}$$

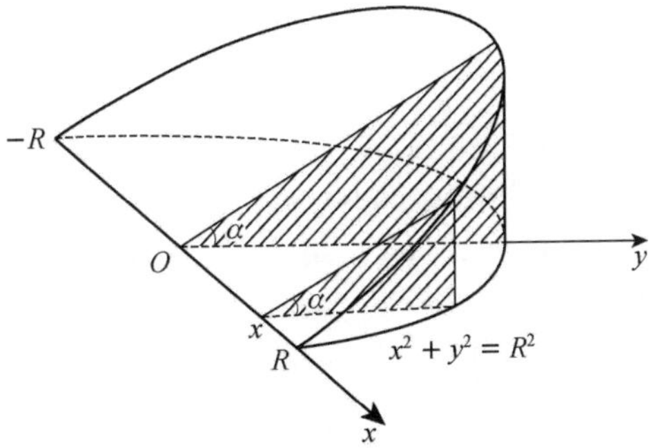

Figure 6.15
图 6.15

Example 8 Suppose that the bottom of a solid is a circle with radius R, and all the sections that are perpendicular to the diameter of the bottom circle are equilateral triangles. Find the volume of the solid.

例 8 求以半径为 R 的圆为底且垂直于底圆直径的所有截面都是等边三角形的立体的体积.

Solution Suppose that the bottom circle lies in Oxy plane, and the center of the circle is the origin O. The equation of bottom circle is $x^2+y^2=R^2$. We make a plane that passes through x which is on the x-axis ($-R<x<R$) and is perpendicular to x-axis, then the corresponding section is an equilateral triangle, and the length of its side is $2\sqrt{R^2-x^2}$ and its height is $\sqrt{3}\sqrt{R^2-x^2}$, so the area of the cross section is

$$S(x)=\frac{1}{2}2\sqrt{R^2-x^2}\cdot\sqrt{3}\sqrt{R^2-x^2}$$
$$=\sqrt{3}(R^2-x^2).$$

Thus, the volume of the solid is

$$V=\int_{-R}^{R}\sqrt{3}(R^2-x^2)\mathrm{d}x$$
$$=\frac{4\sqrt{3}}{3}R^3.$$

3) The Arc Length of a Curve on the Plane

● Rectangular Situation

Let the equation of the arc of a curve in rectangular coordinate is

$$y=f(x)\quad(a\leqslant x\leqslant b),$$

and the function $f(x)$ has the first order continuous derivative on the interval $[a,b]$. Now let's find the length of the arc of the curve.

The abscissa x is regarded as integral variable, and it changes on the interval $[a,b]$. For the length of an arc of the curve $y=f(x)$ corresponding to an arbitrary subinterval $[x,x+\mathrm{d}x]$ in the interval $[a,b]$, we can use a short length of the tangent line of the curve at the point $(x,f(x))$ to approximately instead of it, while the corresponding short length on the tangent line is

$$\sqrt{(\mathrm{d}x)^2+(\mathrm{d}y)^2}=\sqrt{1+y'^2}\mathrm{d}x.$$

Thus, the arc length element is (that is the differential of the arc)

$$\mathrm{d}s=\sqrt{1+y'^2}\mathrm{d}x.$$

Taking $\sqrt{1+y'^2}\,\mathrm{d}x$ as the integrand expression, finding the integration on the closed interval $[a,b]$, we can get the arc length, that is

解 设底圆所在的平面为 Oxy 平面，圆心为原点 O，则底圆的方程为

$$x^2+y^2=R^2.$$

过 x 轴上的点 $x(-R<x<R)$ 作垂直于 x 轴的平面，其所对应的截面是等边三角形，它的边长为 $2\sqrt{R^2-x^2}$，高为 $\sqrt{3}\sqrt{R^2-x^2}$，所以这一截面的面积为

$$S(x)=\frac{1}{2}2\sqrt{R^2-x^2}\cdot\sqrt{3}\sqrt{R^2-x^2}$$
$$=\sqrt{3}(R^2-x^2).$$

于是所求的立体体积为

$$V=\int_{-R}^{R}\sqrt{3}(R^2-x^2)\mathrm{d}x$$
$$=\frac{4\sqrt{3}}{3}R^3.$$

3) 平面曲线的弧长

● 直角坐标情形

设曲线弧由直角坐标方程

$$y=f(x)\quad(a\leqslant x\leqslant b)$$

给出，其中函数 $f(x)$ 在区间 $[a,b]$ 上具有一阶连续导数. 现在来计算这曲线弧的长度.

取横坐标 x 为积分变量，它的变化区间为 $[a,b]$. 曲线 $y=f(x)$ 上相应于 $[a,b]$ 上任一小区间 $[x,x+\mathrm{d}x]$ 的一段弧的长度，可以用该曲线在点 $(x,f(x))$ 处的切线上相应的一小段的长度来近似代替，而切线上相应的这一小段的长度为

$$\sqrt{(\mathrm{d}x)^2+(\mathrm{d}y)^2}=\sqrt{1+y'^2}\mathrm{d}x,$$

从而得弧长微元（即弧微分）

$$\mathrm{d}s=\sqrt{1+y'^2}\mathrm{d}x.$$

以 $\sqrt{1+y'^2}\,\mathrm{d}x$ 为被积表达式，在闭区间 $[a,b]$ 上作定积分，便得所求的弧长为

$$s = \int_a^b \sqrt{1+y'^2}\,dx. \qquad (6.6)$$

Example 9 Find the arc length of the curve $y = \dfrac{2}{3}x^{\frac{3}{2}}$ as $x \in (a,b)$.

Solution Since $y' = x^{\frac{1}{2}}$, then the arc length element is
$$ds = \sqrt{1+y'^2}\,dx = \sqrt{1+x}\,dx.$$
Thus, the arc length is
$$\begin{aligned} s &= \int_a^b \sqrt{1+x}\,dx \\ &= \left[\frac{2}{3}(1+x)^{\frac{3}{2}}\right]_a^b \\ &= \frac{2}{3}\left[(1+b)^{\frac{3}{2}} - (1+a)^{\frac{3}{2}}\right]. \end{aligned}$$

- Polar Situation

Let the equation of the arc of a curve in rectangular coordinate is
$$\rho = \rho(\theta) \quad (\alpha \leqslant \theta \leqslant \beta),$$
and the function $\rho(\theta)$ has continuous derivative on the interval $[\alpha,\beta]$. According to the relationship between the rectangular coordinates and the polar coordinates, we can get
$$x = \rho(\theta)\cos\theta, \quad y = \rho(\theta)\sin\theta \quad (\alpha \leqslant \theta \leqslant \beta).$$
Then, the arc length element is
$$\begin{aligned} ds &= \sqrt{x'^2(\theta)+y'^2(\theta)}\,d\theta \\ &= \sqrt{\rho^2(\theta)+\rho'^2(\theta)}\,d\theta. \end{aligned}$$
Thus, the arc length is
$$s = \int_\alpha^\beta \sqrt{\rho^2(\theta)+\rho'^2(\theta)}\,d\theta. \qquad (6.7)$$

Example 10 Find the arc length of the Archimedean spiral $\rho = a\theta$ as $\theta \in (0, 2\pi)$ (Figure 6.16).

Solution The arc length element is
$$ds = \sqrt{a^2\theta^2 + a^2}\,d\theta = a\sqrt{1+\theta^2}\,d\theta.$$
Thus, the arc length is
$$\begin{aligned} s &= \int_0^{2\pi} a\sqrt{1+\theta^2}\,d\theta \\ &= \frac{a}{2}\left[2\pi\sqrt{1+4\pi^2} + \ln(2\pi+\sqrt{1+4\pi^2})\right]. \end{aligned}$$

$$s = \int_a^b \sqrt{1+y'^2}\,dx. \qquad (6.6)$$

例 9 计算曲线 $y = \dfrac{2}{3}x^{\frac{3}{2}}$ 在 $x \in (a,b)$ 一段的长度.

解 由于 $y' = x^{\frac{1}{2}}$，从而弧长微元为
$$ds = \sqrt{1+y'^2}\,dx = \sqrt{1+x}\,dx,$$
因此所求的弧长为
$$\begin{aligned} s &= \int_a^b \sqrt{1+x}\,dx \\ &= \left[\frac{2}{3}(1+x)^{\frac{3}{2}}\right]_a^b \\ &= \frac{2}{3}\left[(1+b)^{\frac{3}{2}} - (1+a)^{\frac{3}{2}}\right]. \end{aligned}$$

- 极坐标情形

设曲线弧由极坐标方程
$$\rho = \rho(\theta) \quad (\alpha \leqslant \theta \leqslant \beta)$$
给出，函数 $\rho(\theta)$ 在区间 $[\alpha,\beta]$ 上具有连续导数. 由直角坐标与极坐标的关系可得
$$x = \rho(\theta)\cos\theta, \quad y = \rho(\theta)\sin\theta \quad (\alpha \leqslant \theta \leqslant \beta).$$
于是得弧长微元为
$$\begin{aligned} ds &= \sqrt{x'^2(\theta)+y'^2(\theta)}\,d\theta \\ &= \sqrt{\rho^2(\theta)+\rho'^2(\theta)}\,d\theta, \end{aligned}$$
从而所求的弧长为
$$s = \int_\alpha^\beta \sqrt{\rho^2(\theta)+\rho'^2(\theta)}\,d\theta. \qquad (6.7)$$

例 10 求阿基米德螺线 $\rho = a\theta\ (a>0)$ 在 $\theta \in (0,2\pi)$ 一段的弧长 (图 6.16).

解 弧长微元为
$$ds = \sqrt{a^2\theta^2+a^2}\,d\theta = a\sqrt{1+\theta^2}\,d\theta,$$
于是所求的弧长为
$$\begin{aligned} s &= \int_0^{2\pi} a\sqrt{1+\theta^2}\,d\theta \\ &= \frac{a}{2}\left[2\pi\sqrt{1+4\pi^2} + \ln(2\pi+\sqrt{1+4\pi^2})\right]. \end{aligned}$$

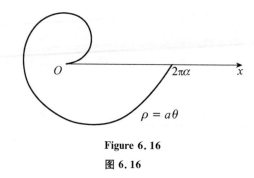

Figure 6.16
图 6.16

3. Applications in Economics

Marginal cost is the change in the total cost that arises when the quantity produced is incremented by one unit. Marginal revenue is the additional revenue that will be generated by increasing product sales by one unit. It can also be described as the unit revenue the last item sold has generated. Marginal profit is the difference between the marginal revenue and the marginal cost. In Economics, if we have known the marginal cost, the marginal revenue and the marginal profit, we can get the total cost function, the total return function and the total profit function:

(1) If the marginal cost is $C'(x)$, then the total cost function is
$$C(x) = \int_0^x C'(x)dx + C_0,$$
where C_0 is a fixed cost;

(2) If the marginal revenue is $R'(x)$, then the total return function is
$$R(x) = \int_0^x R'(x)dx;$$

(3) If the marginal profit is $R'(x) - C'(x)$, then the total profit function is
$$L(x) = \int_0^x [R'(x) - C'(x)]dx - C_0.$$

Example 11 When the amount of some product is x (unit: 100 pieces), the marginal cost and the marginal revenue (unit: 10 thousand yuan) are respectively
$$MC = C'(x) = 2, \quad MR = R'(x) = 10 - 2x.$$

3. 在经济中的应用

边际成本表示当产量增加 1 个单位时, 总成本增加多少; 边际收益指如果再多销售 1 个单位的产品将会得到的收益, 或目前最后卖出的 1 个单位的产品所得到的收益; 边际利润指边际收益与边际成本的差额. 在经济学中, 已知经济函数的边际成本、边际收益和边际利润, 我们可求得总成本函数、总收益函数和总利润函数:

(1) 若边际成本为 $C'(x)$, 则总成本函数为
$$C(x) = \int_0^x C'(x)dx + C_0,$$
其中 C_0 为固定成本;

(2) 若边际收益为 $R'(x)$, 则总收益函数为
$$R(x) = \int_0^x R'(x)dx;$$

(3) 若边际利润为 $R'(x) - C'(x)$, 则总利润函数为
$$L(x) = \int_0^x [R'(x) - C'(x)]dx - C_0.$$

例 11 当某产品的产量为 x(单位: 百件)时, 其边际成本和边际收益(单位: 万元)分别为
$$MC = C'(x) = 2, \quad MR = R'(x) = 10 - 2x.$$

(1) How much the amount of the product will be when it maximize the total profit?

(2) How much will the profit decrease when the amount of the product which maximizes the total profit increases by two hundred?

Solution (1) Since
$$L'(x) = R'(x) - C'(x)$$
$$= 10 - 2x - 2 = 8 - 2x,$$
let $L'(x) = 0$, we can get $x = 4$. That is, 400 pieces of product maximize profit.

(2) Since
$$L(x) = \int_0^x L'(x)\mathrm{d}x$$
$$= \int_0^x (8 - 2x)\mathrm{d}x$$
$$= 8x - x^2,$$
we can get
$$L(4+2) - L(4) = 12 - 16 = -4.$$
That is, when the amount of the product which maximize the total profit increases by two hundred, the profit will reduce by 40 thousand yuan.

4. Application in Physics

Definite integral has a wide range of applications in Physics. Here, we will make a brief introduction only with two examples, i. e., the spring doing work and the center of mass of a line.

Example 12 If the natural length of a spring is 0.3 m, and if it takes a force of 12 N to keep it extended 0.04 m, find the work done in stretching the spring from its natural length to length of 0.4 m.

Solution From Hooke's Law, the force $F(x)$ required to keep the spring stretched x meter is given by $F(x) = kx$ (k is the coefficient of stiffness). We note that $0.04k = 12$, or $k = 300$, and so
$$F(x) = 300x.$$
When the length of the spring is 0.4 m, $x = 0.1$. The element of the spring doing work is $300x\mathrm{d}x$, therefore, the work done in stretching the spring is
$$W = \int_0^{0.1} 300x\mathrm{d}x = [150x^2]_0^{0.1}$$
$$= 1.5 \text{ (unit: J)}.$$

Example 13 The density $\delta(x)$ of a wire at the point x centimeters from one end is given by $\delta(x) = 3x^2$ (unit: g/cm). Find the center of mass of the piece between $x = 0$ and $x = 10$.

Solution Divide this piece of wire into n equal parts. When n is sufficiently large, the center of mass is

$$\overline{x} \approx \sum_{i=1}^{n} m_i \Delta x / m,$$

where m_i is the mass of part i, m is the total mass. Because mass element is $dm = \delta(x)dx$, when $n \to \infty$, then

$$\overline{x} = \frac{\int_0^{10} x\delta(x)dx}{\int_0^{10} \delta(x)dx}.$$

Thus, we get

$$\overline{x} = \frac{\int_0^{10} x \cdot 3x^2 dx}{\int_0^{10} 3x^2 dx} = \frac{\left[\frac{3x^4}{4}\right]_0^{10}}{\left[x^3\right]_0^{10}}$$

$$= \frac{7500}{1000} = 7.5.$$

So the center of mass of this piece of wire is at the point $x = 7.5$.

Exercises 6

1. Compute $\int_{-1}^{1} (x^2 + 1) dx$ by definition of definite integral.

2. Applying geometrical significance of definite integral to compute the following definite integrals:

(1) $\int_{-1/2}^{1} (2x + 1) dx$;

(2) $\int_0^3 \sqrt{9 - x^2} dx$;

(3) $\int_{-\pi}^{\pi} \sin x dx$.

3. Known $\int_{-1}^{2} f(x)dx = 5$, $\int_{2}^{5} f(x)dx = 4$, $\int_{-1}^{2} g(x)dx = 3$, compute the following definite integrals:

(1) $\int_{-1}^{2} 6f(x)\mathrm{d}x$;

(2) $\int_{-1}^{5} f(x)\mathrm{d}x$;

(3) $\int_{-1}^{2} \frac{1}{3}[4f(x)-5g(x)]\mathrm{d}x$;

(4) $\int_{5}^{2} f(x)\mathrm{d}x$.

4. Compute the following definite integrals:

(1) $\int_{1}^{2}(x^2+3x+4)\mathrm{d}x$;

(2) $\int_{1}^{3}\left(x^4-\frac{2}{x}\right)\mathrm{d}x$;

(3) $\int_{1}^{4}\left(\sqrt{x}-\frac{1}{\sqrt[3]{x}}\right)\mathrm{d}x$;

(4) $\int_{1}^{\sqrt{3}} \frac{\mathrm{d}x}{1+x^2}$;

(5) $\int_{-\sqrt{3}/2}^{\sqrt{3}/2} \frac{\mathrm{d}x}{\sqrt{1-x^2}}$.

5. Applying the L'Hospital rule, compute the following limits:

(1) $\lim\limits_{x\to 0} \dfrac{\left(\int_{0}^{x} e^{t^2}\mathrm{d}t\right)^2}{\int_{0}^{x} te^{t^2}\mathrm{d}t}$;

(2) $\lim\limits_{x\to 0} \dfrac{\int_{0}^{x} \sin t^2 \ln(1+t)\mathrm{d}t}{x^3 \tan(\sqrt{1+x}-1)}$.

6. Applying integration by substitution, compute the following indefinite integrals:

(1) $\int_{\pi/3}^{\pi} \cos\left(x+\frac{\pi}{3}\right)\mathrm{d}x$;

(2) $\int_{-1}^{2} \frac{\mathrm{d}x}{(2+3x)^3}$;

(3) $\int_{0}^{\pi/4} \sin^4 x\cos x\mathrm{d}x$;

(4) $\int_{\pi/6}^{\pi/2} \sin^2\theta\mathrm{d}\theta$;

(5) $\int_{0}^{3} \sqrt{9-x^2}\mathrm{d}x$.

7. Applying partial integral method, find the following definite integrals:

(1) $\int_{0}^{\pi/2} x\cos 4x\mathrm{d}x$;

(2) $\int_{\pi/4}^{\pi/3} \frac{x}{\cos^2 x}\mathrm{d}x$;

(1) $\int_{-1}^{2} 6f(x)\mathrm{d}x$;

(2) $\int_{-1}^{5} f(x)\mathrm{d}x$;

(3) $\int_{-1}^{2} \frac{1}{3}[4f(x)-5g(x)]\mathrm{d}x$;

(4) $\int_{5}^{2} f(x)\mathrm{d}x$.

4. 计算下列定积分:

(1) $\int_{1}^{2}(x^2+3x+4)\mathrm{d}x$;

(2) $\int_{1}^{3}\left(x^4-\frac{2}{x}\right)\mathrm{d}x$;

(3) $\int_{1}^{4}\left(\sqrt{x}-\frac{1}{\sqrt[3]{x}}\right)\mathrm{d}x$;

(4) $\int_{1}^{\sqrt{3}} \frac{\mathrm{d}x}{1+x^2}$;

(5) $\int_{-\sqrt{3}/2}^{\sqrt{3}/2} \frac{\mathrm{d}x}{\sqrt{1-x^2}}$.

5. 用洛必达法则计算下列极限:

(1) $\lim\limits_{x\to 0} \dfrac{\left(\int_{0}^{x} e^{t^2}\mathrm{d}t\right)^2}{\int_{0}^{x} te^{t^2}\mathrm{d}t}$;

(2) $\lim\limits_{x\to 0} \dfrac{\int_{0}^{x} \sin t^2 \ln(1+t)\mathrm{d}t}{x^3 \tan(\sqrt{1+x}-1)}$.

6. 用换元积分法求下列定积分:

(1) $\int_{\pi/3}^{\pi} \cos\left(x+\frac{\pi}{3}\right)\mathrm{d}x$;

(2) $\int_{-1}^{2} \frac{\mathrm{d}x}{(2+3x)^3}$;

(3) $\int_{0}^{\pi/4} \sin^4 x\cos x\mathrm{d}x$;

(4) $\int_{\pi/6}^{\pi/2} \sin^2\theta\mathrm{d}\theta$;

(5) $\int_{0}^{3} \sqrt{9-x^2}\mathrm{d}x$.

7. 用分部积分法求下列定积分:

(1) $\int_{0}^{\pi/2} x\cos 4x\mathrm{d}x$;

(2) $\int_{\pi/4}^{\pi/3} \frac{x}{\cos^2 x}\mathrm{d}x$;

(3) $\int_1^8 \dfrac{\ln x}{\sqrt[3]{x}} dx$.

8. Compute the following improper integrals:

(1) $\int_1^{+\infty} \dfrac{dx}{x^3}$;

(2) $\int_1^{+\infty} \dfrac{dx}{\sqrt{x}}$;

(3) $\int_0^{+\infty} \dfrac{dx}{e^{1+x}+e^{3-x}}$;

(4) $\int_0^{+\infty} x\,e^{-x^2} dx$;

(5) $\int_0^2 \dfrac{x\,dx}{\sqrt{4-x^2}}$;

(6) $\int_0^2 \dfrac{dx}{(2-x)^2}$;

(7) $\int_1^2 \dfrac{dx}{x\ln x}$.

9. Applying definite integral to compute the area of the region enclosed by the lines $y=x$, $x=4$ and x-axis.

10. Compute the area of the region enclosed by the parabola $y=x^2$ and the line $y-2x=0$.

11. Compute the area of the region enclosed by the curved line $y=-x^3+3x^2-2x$ and x-axis.

12. Compute the area of the region enclosed by the curve $y=x+\dfrac{1}{x}$, the lines $x=2$ and $y=2$.

13. Compute the volume of the solid which obtained by rotating about x-axis the region enclosed by the curves $y=e^x$, $y=\sin x$ and the lines $x=0, x=1$.

14. Compute the length of the catenary $y=\dfrac{1}{2}(e^x+e^{-x})$ as $x\in[0,t]$.

15. When the amount of some product is x (unit: 100 pieces), the marginal cost and the marginal revenue (unit: 10 thousand yuan) are

$MC=C'(x)=1.5$, $MR=R'(x)=8-2x$.

(1) How much the amount of the product will be when the total profit gets a maximum?

(2) How much will the profit decrease, when the amount of the product which maximizes the total profit increases by four hundred?

(3) $\int_1^8 \dfrac{\ln x}{\sqrt[3]{x}} dx$.

8. 计算下列反常积分:

(1) $\int_1^{+\infty} \dfrac{dx}{x^3}$;

(2) $\int_1^{+\infty} \dfrac{dx}{\sqrt{x}}$;

(3) $\int_0^{+\infty} \dfrac{dx}{e^{1+x}+e^{3-x}}$;

(4) $\int_0^{+\infty} x\,e^{-x^2} dx$;

(5) $\int_0^2 \dfrac{x\,dx}{\sqrt{4-x^2}}$;

(6) $\int_0^2 \dfrac{dx}{(2-x)^2}$;

(7) $\int_1^2 \dfrac{dx}{x\ln x}$.

9. 用定积分计算由直线 $y=x, x=4$ 及 x 轴所围成的平面图形的面积.

10. 计算由抛物线 $y=x^2$ 及直线 $y-2x=0$ 所围成的平面图形的面积.

11. 求曲线 $y=-x^3+3x^2-2x$ 与 x 轴所围成的平面图形的面积.

12. 求由曲线 $y=x+\dfrac{1}{x}$ 与直线 $x=2$ 及 $y=2$ 所围成的平面图形的面积.

13. 求由曲线 $y=e^x, y=\sin x$ 与直线 $x=0, x=1$ 所围成的平面图形绕 x 轴旋转所成立体的体积.

14. 计算悬链线 $y=\dfrac{1}{2}(e^x+e^{-x})$ 在 $[0,t]$ 一段的弧长.

15. 当某产品的产量为 x(单位:百件)时,其边际成本和边际收益(单位:万元)分别为

$MC=C'(x)=1.5$, $MR=R'(x)=8-2x$,

(1) 产量为多少时总利润最大?

(2) 当产量从最大利润时的产量再增加 4 百件时,利润将减少多少?